Adult Learning in the Digital Age

Learning with technology is seen by many countries as a way of improving their collective human 'capital', and of offering opportunities to those currently excluded from mainstream education. But in reality how do adults learn to use, and learn by using, information and communications technologies (ICTs) such as the computer, the internet and digital television? To what extent are we witnessing a technology-based learning revolution?

This illuminating and engaging book provides hard facts on a topic that is all too often clouded by rhetoric as it describes the different ways in which adults interact with ICTs for learning at home, work and in the wider community. Written by acknowledged experts in the field, the valuable research findings will spark fresh debate on:

- Why ICTs are believed capable of effecting positive change in adult learning
- The drawbacks and limits of ICT in adult education
- How people can be made into lifelong learners
- What people use ICT for in the home, work and community
- The wider social, economic, cultural and political realities of the information age and the learning society.

Adult Learning in the Digital Age is based on one of the first large-scale academic research projects in the area. Its rich and detailed findings are used to generate practical recommendations for policy-makers, practitioners and future researchers interested in the use of new technology and the development of the information society.

Neil Selwyn is Senior Lecturer at the School of Social Sciences, University of Cardiff. **Stephen Gorard** is Research Professor at the Department of Educational Studies, University of York. **John Furlong** is the Director of the Department of Educational Studies, University of Oxford.

Adult Learning in the Digital Age

Information technology and the learning society

Neil Selwyn, Stephen Gorard and John Furlong

Routledge
Taylor & Francis Group

LONDON AND NEW YORK

First published 2006
by Routledge
2 Park Square, Milton Park, Abingdon, Oxon OX14 4RN

Simultaneously published in the USA and Canada
by Routledge
270 Madison Ave, New York, NY 10016

Routledge is an imprint of the Taylor & Francis Group

Typeset in Goudy by
HWA Text and Data Management, Tunbridge Wells
Printed and bound in Great Britain by
The Cromwell Press, Trowbridge, Wiltshire

British Library Cataloguing in Publication Data
A catalogue record for this book is available from the British Library

Library of Congress Cataloging in Publication Data
Selwyn, Neil
    Adult learning in the digital age: information technology and the
learning society / Neil Selwyn, Stephen Gorard and John Furlong.
        p.   cm.
    Includes bibliographical references and index.
    1. Adult learning.   2. Adult education – Computer-assisted instruction.
3. Information technology.   4. Educational technology.   I. Gorard, Stephen.
    II. Furlong, John, 1947–   III. Title.
LC5225.L42S48 2005

374–dc22   20500730

ISBN 0–415–35698–9 (hbk)
ISBN 0–415–35699–7 (pbk)

Contents

Illustrations

Figures

Tables

Preface

I've learnt to use computers properly now ... but I've got nothing to use them properly for.

(male, 35 years old)

Background

The digital age presents many challenges for those in education and government. The need for the whole population to be able to access and use new technologies is seen as a crucial first step in establishing a skilled workforce and empowered citizenry for the twenty-first century. The potential for information and communications technologies (ICTs) to allow people to learn at all stages of life has been seized upon as a ready means of establishing inclusive 'learning societies'. New technologies are purportedly ushering in a new age of lifelong learning, which can be centred around the individual learner while remaining cost-effective for the educational provider. Grand claims have also been made about technology-based lifelong learning underpinning countries' competitiveness in a global knowledge economy. It is of little surprise, then, that governments around the world have spent the past ten years setting targets and implementing multi-billion dollar policies aimed at encouraging adults to live, work, and learn with the support of ICTs.

The political convergence of the high-tech and high-skill agendas is the latest example of politicians and governments coming to terms with the demands of the post-industrial 'information society'. As such, striving to establish a technology-based learning society is a logical extension of the e-commerce boom of the late 1990s and the ongoing 'e-government' transformation of the public sector. We are living in an age where technology is seen as more than capable of taking care of matters like education. When the prospect of school children taking their public examinations via mobile phone is being seriously considered by educational authorities, it is sensible enough to assume that mainstream ICTs such as the computer and the internet can be harnessed for the straightforward task of supporting adults to take part in education. Indeed, the view that technology-based adult education is a 'good thing' is now such a widely-held belief that few educational commentators bother to state it. They prefer, instead, to talk in terms of when, not if, education will be 'transformed' by technology. For many of these 'cyber-gurus' and 'e-evangelists' of the 1980s and 1990s, the digital wars have been won. Apparently, we now live in a society where virtually everyone has access to computers, mobile phones, digital television and the internet, and where many people are already using these technologies for the basics of life such as grocery shopping and paying their taxes. Living our everyday lives online, many of us will also hopefully be persuaded to make use of the same technologies to learn things for pleasure or gain new skills for the workplace. The beauty of ICTs, so the received wisdom goes, is that they give everyone the opportunity to learn and can therefore 'create' lifelong learners of us all:

> it seems technology is assuming a new role for many in the UK ... technology [is] doing more than just enabling us to do more of what we already love to do ... 41 percent of those polled stated that the internet was responsible for getting them involved in *new* activities, sports, hobbies and classes.
>
> (ICM 2004: 26)

Many educationalists also appear to have become ensnared in the seductive rhetoric of the information society, but they are presuming an ubiquity of technology of a kind which simply does not exist in early twenty-first century Britain. These utopian accounts of an educational near-future would be more credible were it not for the fact that as a society 'we' do not all have access to ICTs. Indeed, 'we' do not all have an immediate need to live our everyday lives through new technologies, and 'we' certainly are not all finding the need to take part in any education or training as a result. It is, perhaps, too easy for those who do use ICTs frequently to project a universal importance and sense of empowerment which many others do not experience. But there is little objective reason to expect the present-day 'knowledge economy' and 'information society' to be any less divided and unequal than economies and societies have always been. In short, we should not let the allure of new technologies cloud our critical faculties.

Yet it is difficult to maintain a critical objectivity when technology appears to be having such a profound, beneficial impact on so many aspects of the education process. The pedagogical application of new technologies can be truly breathtaking – especially for those who worked in education before the days of the microchip. From digitised 'e-books' to video-conferencing, new technologies are being used to amazing educational effect in schools, colleges, universities, workplaces and homes. Once one has experienced a technology like the electronic whiteboard or PDA used to its full educational potential it is difficult not to believe that the key to establishing a society based around learning for all lies in the application of such new technology.

The one stumbling block to this argument is the under-whelming evidence of any *sustained* technological transformation of education 'on the ground'. ICT-based learning is not actually an integral element of most educational settings that we are likely to encounter. Although there are many fine examples of 'best practice' up and down the country, video-conferencing and e-books are not the norm in most classrooms or lecture theatres in the UK. Similarly, in terms of adult learning, considerably more people use their interactive digital television sets to watch feature films and football matches than to learn for pleasure or study for a qualification. In short, there has always been a considerable discrepancy between the rhetoric and the reality of the notion of an ICT-driven 'learning society' – a disparity which social researchers have been curiously slow to engage with. Indeed, the initial impetus for this book was our observation that the topic of adult learning in the digital age was being much talked about but little investigated. Given the vast amounts of money, resources and effort being invested we were compelled to find out more about how close the UK was to becoming a technology-based 'learning society', and what problems may be faced along the way. Within the existing literature we found that key questions of *who* is using ICTs for *what* purposes and with *what* outcomes remained largely unanswered. Academic understanding of who was *not* using ICT and why they were not doing so was even vaguer. Beyond recognising the promise and potential of new technologies, mapping how ICTs and ICT-based learning fit in with the everyday lives of adults was, and still is, a vital task for the research community.

With hindsight, addressing these questions was not as straightforward as we first assumed. It is easy to imagine from the high-profile funding and branding of 'e-learning' initiatives such as learndirect and UK Online that ICT-based adult learning is a mainstream, commonplace activity. The reality, even in a self-proclaimed world-leading nation such as the UK, is somewhat different, as these field notes taken on an initial visit to one of our research sites reflect:

> At just after three o'clock in the afternoon we drive down into the centre of Radstock – a small town to the south of Bath. Parking by the Mining Museum with its restored pit wheel we notice 'Wired World' – a three storey grey brick building advertising itself as a cybercafé. Entering the building there is a bank of eight PCs all being used by teenagers to play Doom-type games, a bar area in the centre and a row of sofas to the left. The owner informs us that the cybercafé does not offer access to the internet or email. It is generally used by youngsters to play games against each other, although they do run a session on Monday evenings 'for guys like us' to play games. We assume that by 'guys like us' the owner means those over the age of eighteen rather than anything more prurient. He then tells us that both the local library and local FE college have drop-in internet facilities. The library, which proves to be about three minutes walk away, does seem to offer internet access and colour printing from a PC. However, as it is not open on a Friday we cannot go in to investigate further. Driving two minutes up the road, the local FE college looks more promising. The sign at the entrance to the college boasts 'Internet Café – Open to the Public'. Having parked up, three of the seven options on

the main sign are for 'IT Suite', 'Internet Café' and 'IT Training Centre'. Following these signs leads us to a room with a metal shutter that appears to have once been an internet café. Moving on down the empty corridor we come across a computer lab with a cleaner and a solitary student. The student (who is French) tells us that there used to be an internet café but he thinks that it has moved into town. He suggests we try the main college reception for more information. The reception is a very spacious and well furnished area with three stands of 'learndirect' leaflets placed on either side of the room. The receptionist confirms that the internet café on site is defunct but that she thinks the new outreach centre that the college has set up in the middle of town 'may offer internet access'. She rings the outreach centre for us and then walks into another room leaving us with the telephone. No-one answers and after two minutes of ringing we give up as there seems to be no answer phone. It is three-thirty in the afternoon and had we been regular 'punters' looking for ICT-based learning we would have probably given up long ago. On a rainy day in Radstock the digital age feels a long way away.

(Fieldnotes 8/3/2002)

Despite this inauspicious start we persevered. This book is the culmination of our subsequent two and a half year study of adult learning and technology – one of the first large-scale social research projects to focus specifically on ICTs and adult learning. We hope that it brings a powerful empirical dimension to the existing debate as well as provoking a more realistic discussion within the education and technology communities about what can, and what cannot, be achieved through the use of ICTs. Our findings are based on a detailed survey of 1,001 adults in England and Wales with a further 'booster' sample of 100 users of public ICT centres, follow-up interviews with a subset of 100 of the respondents, and year-long case studies of 25 technology-using individuals and their families. Through a set of rich and often thought-provoking pictures of how people are using ICTs in their day-to-day lives, the book highlights the 'messy' realities behind the rhetoric of the 'le@rning society'. We have also taken the time to tell the stories of the non-participants – a substantial group not only excluded from learning and technology but usually also excluded from research about their exclusion from education. From this, we have produced a set of powerful but stark recommendations for educationalists, politicians and policymakers. We hope that our work will be used both practically and theoretically by those charged with creating an ICT-based adult learning sector.

Organisation of the book

We present our work in eleven substantive chapters, starting with two chapters which consider the arguments for and against the prevailing 'revolutionary' rhetoric of ICT-based adult learning. Chapter 1 outlines the basis for the current excitement and enthusiasm surrounding ICT-based lifelong learning. It discusses why so many believe passionately that ICTs are capable of effecting radical educational change. In so doing, the chapter maps the long-held belief in the inevitable technological transformation of adult learning. It draws on a wide range of 'techno-enthusiasts' and commentators – from Manual Castells to Nicholas Negroponte – and presents the compelling case for ICT and adult education within the wider political imperatives of the information society and knowledge economy.

Chapter 2 provides a balance to the preceding optimism by gathering together some of the key critical commentaries on ICTs and lifelong learning from authors and studies often overlooked in the education and technology literatures. Here we see that for all the perceived benefits of ICT there are a host of caveats, potential drawbacks and unresolved problems. The chapter contends that, in order to use ICT in effective ways, educationalists need to develop more sophisticated understandings of the enduring 'digital divides' of educational participation and ICT use, as well as facing up to the educational and pedagogical limitations of ICT.

Having reviewed the existing debate in the area, Chapter 3 goes on to introduce our *Adult Learning@Home* research project, and outline the research principles we adopted in developing an empirically rigorous approach to researching adult learning and technology. For the methodologically-minded reader the second half of the chapter presents descriptions and justifications of the research design, methods and sampling strategies used in the project. Less methodologically-interested readers will also find this section of value because it describes the twelve communities in which the research took place, and so provides a key context for the findings which follow.

Chapters 4 and 5 present the main findings from our household survey. Chapter 4 describes the varying patterns of participation in lifelong learning evident in the data, showing that over a third of all adults reported no further *formal* education or training of any kind after reaching compulsory school-leaving-age. We find that the actual patterns of adult participation in education are predictable to a large extent from individuals' early life experiences. The key indicators include age, sex, family background, and initial schooling – all of which are set very early in life. As a result of these factors individuals can create for themselves a kind of 'learner identity' either inimicable to or disposed towards further study. In the former case, the prospect of learning can become a burden rather than an investment. This leads us to suggest that universal theories to describe the causes of participation in adult learning, such as human capital theory, cannot be sustained. As our survey data show, this has significant implications for the notion of overcoming barriers to access via technology.

Expanding this last point, Chapter 5 examines the survey data in terms of the learning activities that ICTs were actually being used for throughout our population of study. Here we find that our respondents report education and learning to be minority activities when using a computer, most commonly taking the form of informal learning at home. Any educative use of ICTs that does occur appears to be patterned by a number of social factors. In particular whether or not an individual reported the use of ICTs for educative purposes can be predicted by the same indicators that predict other kinds of learning as well, such as age, initial education, occupational class and area of residence. That said, informal learning taking place through leisure interests, work, and day-to-day life is found to be at least as important as formal learning provision, particularly for procedural knowledge about the use of ICTs. Drawing on our follow-up interview data we then go on to highlight the range of informal learning that *is* taking place, and how ICT both facilitates and suppresses such learning opportunities.

Chapter 6 is the first of four chapters presenting key findings from our follow-up interviews and longitudinal case studies. It considers ICT-based learning in the social context of the home, and explores how new technologies and ICT-based learning are interwoven with domestic life. Although learning and education were found to be prominent in the 'repertoires of official reasons' which adults invoked to introduce the technology into the household, subsequent educational use of computers was contingent on a host of social, economic and cultural reasons. In particular, we detail how computers in the home were largely allowing people to engage further in the types of learning which they were already engaged in. Thus, where respondents were found to be using the internet avidly for informal learning or self-education, they also reported having previously done so with books, magazines, television programmes, friends and neighbours. In most cases, these learners continued to learn in such 'traditional' ways alongside ICTs, suggesting that new technologies were fitting alongside these existing technologies and techniques of learning rather than supplanting them.

Chapter 7 takes a similar approach to examining adults' ICT use in the workplace and the learning which this entails. A range of findings are highlighted and discussed. For example, for a minority of our respondents work and home had become intertwined in ways that made them almost indistinguishable. In most cases, ICTs had been central to that transformation – especially in terms of home-working, and home-learning for work. In trying to explain this transfer of learning, one obvious finding is that many adults had very different opportunities to engage with ICTs through the processes of their work and in the physical context of their workplace. Whether our respondents had had the *opportunity* to develop their skills in relation to ICT through work therefore varied considerably depending on their age (some who were retired were simply too old to have had this experience), nature of employment and the speed with which their particular employer had introduced ICTs into the workplace.

Given the current political emphasis on the 'community' provision of ICT access and training, Chapter 8 goes on to examine the effectiveness of public sites in facilitating adults' engagement with ICT and ICT-based learning. In particular it combines our booster survey of 100 people using drop-in centres, with interview and case study data to explore how and why respondents were making use (or not making use) of public ICT provision and with what outcomes. These data suggest that whilst public ICT sites were leading to valuable outcomes for *some* individuals in terms of technological confidence and competence, participation in education and training and the development of employment-related skills, this was generalisable only to a minority of our respondents. The more common picture of non-engagement appeared to be prompted by a number of reasons including the 'exclusive' nature of public ICT sites and the types of ICT usage they supported and, significantly, a general lack of any need to use ICT in day-to-day life.

The final empirical chapter of the book – Chapter 9 – reflects on how our respondents had learnt to use computers. Through this specific example of the social complexities of what happens when adults actually learn with technology we highlight a number of pertinent issues. For instance, successful and effective use of ICT would appear not merely to be about 'having' or 'not having' formal access to technologies and technology-learning opportunities, but also the scope and intensity of the relationships that people develop with technologies and the nature of what they do with them. Crucially, the boundaries between different contexts of ICT use were often blurred, or even fully integrated. There was also a considerable blurring of formal and informal learning. We found that people were collecting or acquiring elements of computer learning from different sources, in different contexts and to different degrees of success. Learning to use a computer could therefore be described as an ongoing process of 'bricolage' – influenced not only by individuals' material circumstances and their motivational circumstances but also by wider mediating social structures.

Chapter 10 places all of these findings within the wider social, economic, cultural and political realities of the notions of lifelong learning and the information society. A range of overarching conclusions are reached and discussed. Above all our data seem to highlight the 'messiness' of the digital age, and the often mundane and sporadic nature of people's engagement with ICTs. ICT-based learning, despite the 'cyperbole' to the contrary, was not a prominent feature of our data. Where it did appear then technologies such as the computer, internet and digital television were found to be used for informal learning rather than participation in formal education. Much computer-based learning was also found to be highly 'self referential' – involving learning *about* computers rather than *through* them. If used at all for learning then computers were incorporated into adults' existing patterns of life and forms of learning – *supplementing* rather than supplanting 'traditional' forms of learning and *extending* rather than widening adult participation in learning. Similarly, it was clear from all of our data that we should refute the suggestion that technology-using individuals tend to follow a 'general life script' or life-course which is typical to all. Neither would we claim that there are necessarily elements of 'successful' technology-using individuals which can be replicated via public policies or local interventions. Instead, it is clear that people's use and non-use of technologies – and their propensity to use technology for whatever purpose – is often shaped by highly individualised and personal life histories.

Chapter 11 concludes the book by taking these findings, and generating practical recommendations for policymakers, practitioners and future researchers interested in the use of new technology for a learning society. In particular it seeks to move beyond the preceding chapters' somewhat gloomy prognosis of the current situation and discusses what future directions need to be pursued by policymakers, educationalists and other interested stakeholders. The chapter highlights a range of crucial issues such as the role of the state and the private sector in facilitating equitable patterns of ICT use and learning throughout the general population, the need for the readjustment of political expectations of what ICTs can and cannot be used for in the development of learning societies, the need for stakeholders to concentrate on the social as well as technical factors underlying non-participation in education and, finally, a consideration of practical ways in which ICTs can be provided and used to support and facilitate adult learning.

On a final note, one of the less desirable features of the recent rise of ICT in education is the plethora of surrounding terminology and jargon. Whilst we have endeavoured to make distinctions and definitions clear to the reader, it is worth taking the time to introduce three central definitions from the outset of the book. Despite their semantic similarities we use these three phrases to refer to very distinct aspects of adult learning. First, the notion of the *'learning society'* lies at the heart of the book – referring to the political ideal of a society built around its population participating in learning throughout the life-course and the economic and social benefits which result. Second, the much used term of *'e-learning'* is employed in its broadest sense to refer to any type of learning that takes place with the involvement of ICTs (from researching an essay on the internet to taking an online degree as a distance learner). Finally, combining these first two concepts we introduce the notion of the *'le@rning society'* throughout the book to refer to the specific version of the learning society ideal where full participation in education is seen as taking place via ICTs and e-learning. In a sense, then, this book is predicated upon the empirical analysis of the 'le@rning society' thesis.

Acknowledgements

Of course, a two and a half year long research project is not as straightforward as it may appear from our account. Collecting and analysing the data, developing the arguments and reaching the conclusions the book describes has been a long, involved process with others providing help and support along the way. First and foremost the research project upon which this book is based was funded by the Economic and Social Research Council (research grant R000239518). As with all research monographs, whilst we assume responsibility for all errors, inaccuracies and mistakes, we could not have written this book without the assistance of several others. The research presented in this book was carried out by a team of people. Louise Madden worked as a researcher for the duration of the *Adult Learning@Home* project and was involved in the collection of much of the in-depth data. Kate Hubert worked as the project administrator during the data collection stages and had the unenviable task of co-ordinating the fieldwork activities and transcribing many of the interviews. Stephanie Coon and Jamie Lewis joined us as researchers towards the end of the project and assisted with some of the data analysis as well as generating many interesting ideas.

In terms of the writing and production of the book we would like to thank RoutledgeFalmer as well as the anonymous readers for critiquing earlier drafts of the manuscript. Throughout the duration of the project many people have offered help, advice and criticism. These include: Bill Green, Naomi Sargeant, Alan Tuckett, Sonia Liff, Charles Crook, Jeremy Wyatt, Jo Pye, Mike Cushman as well as Jonathan Lee and Dale Hall at ORS. We would also like to thank colleagues past and present for their advice and comments, including: John Fitz, Robert Evans, Trevor Welland, Patrick White, Keri Facer and Tim Rudd.

Particular gratitude is due to those who participated in the project as our 'respondents'. We would like to thank those individuals who gave up the time to answer our questions, show us around their houses and workplaces and put up with our repeated intrusions into their lives. Without these respondents there would have been no project. Although we cannot name any of them they will know who they are.

Neil Selwyn, Stephen Gorard and John Furlong
Cardiff, February 2005

Abbreviations

BBC	British Broadcasting Corporation
CAD	computer aided design
CAL	computer assisted learning
CD	compact disc
CD-ROM	compact disc read only memory
CLAIT	computer literacy and information technology
DfEE	Department for Education and Employment
DfES	Department for Education and Skills
DTV	digital television
DVD	digital versatile disc
ECDL	European computer driving licence
EU	European Union
FE	further education
GCSE	General Certificate of Secondary Education
HE	higher education
HND	higher national diploma
HTML	hypertext mark-up language
IBM	International Business Machines
IBT	integrated business technology
ICT	information and communications technology
IT	information technology
MP3	moving picture experts group audio layer 3
NE	north east
NIACE	National Institute for Adult and Continuing Education
NPA	new political arithmetic
OECD	Organisation for Economic Co-operation and Development
PC	personal computer
PDA	personal digital assistant
RSA	Royal Society of Arts
TV	television
UfI	University for Industry

UK	United Kingdom
US/USA	United States of America
YTS	Youth Training Scheme

Chapter 1

The promise of adult learning in the digital age

Introduction

The information age presents many challenges for those in education and government. The need for the whole population to be able to access and use new technologies such as computers, the internet and digital television is often seen as crucial to establishing a skilled workforce and empowered citizenry for the twenty-first century. The potential of these new technologies to allow people to learn throughout the life-course is also seen as a ready means of establishing developed countries as 'learning societies'. Governments around the world have therefore set targets and developed policies to help all adults to learn, work and live with the support of information and communications technologies (ICTs).

But despite the vast sums of money and effort being directed towards ICT and education we still know little of how close we are to establishing technology-based 'learning societies' and what problems may be faced along the way. Key questions of who is using ICTs for what purposes and with what outcomes remain largely unanswered within the current literature. Our understanding of who is *not* using ICT and why they are not doing so is also vague. Beyond recognising the promise and potential of new technologies, mapping how ICTs and ICT-based learning fit with the everyday lives of adults is a vital task for the research community.

There will be few readers of this book whose lives do not involve the use of information and communications technology. There is, for example, a good chance that a typical reader of an academic or policy-relevant book will have some form of internet access at work or at home. Many will have a mobile phone and perhaps a handheld computing device of some description. In short, the majority of readers should at least notionally be part of the 'information society' – a world in which the 'anytime, anyplace, anypace' reach of telecommunications technologies is transforming people's day-to-day lives. Enthusiastic accounts of the 'power' of new technologies to support and shape our everyday activities proliferate via newspapers, textbooks and television screens. From e-commerce to e-tailing and cybersex to blogging, networked computerised technologies are heralded by some to be as epoch-making as the nineteenth-century industrial revolution. Inspired by grand notions of globalisation and post-modernity some scholars have taken to portraying adults in the early twenty-first century as living in a plentiful post-physical age where all that is solid melts into bits (Negroponte 1995). As William Mitchell (2001) would have us believe, 'e-topia' is but a mouse-click away.

Much of this excitement has been triggered by a recent 'coming of age' of information technologies. Although technologists had long talked of fully interactive and user-centred networked computers, until relatively recently such technologies were conspicuous by their absence on the high street. Yet the last decade or so has witnessed a rapid convergence of computers, telecommunications and broadcasting technologies – resulting in a proliferation of modern-day consumer technologies with remarkable communicative and networking capacities. This new-found availability and affordability has been supported by a 'rebranding' of the term 'information technology' into 'information and communications technology' – a reference to the convergence of technological artefacts such as computers, digital broadcast technologies and mobile phones into platforms all capable of supporting information and communications resources such as the world wide web and email.

ICTs and the political challenges of the information society

From a technical point of view the semantic shift from IT to ICT is an unnecessary one. All the technologies mentioned above can still be described as 'information technologies' – a fact reflected in the failure of 'ICT' to be used widely as a term outside Europe. Yet the phrase is significant in marking a new-found recognition that the growing availability and use of these technologies brings with them a range of important societal and economic implications which cannot be ignored by those involved in the running of countries. As Miles reasons:

> ICTs allow for new rules of the game in all areas of economic activity and [mean] that organisations of all types need to innovate and redefine their objectives in this context. Not all developments will involve the new technologies, but often they will be central. Standing still is rarely an option. Although beleaguered services and individuals may resent being told about the need to adapt, a sea change is taking place.
>
> (Miles 1996: 51)

Governments and those involved in public services therefore find themselves tackling the real-life concern of how ICTs may be best shaped for contemporary society. Although the 'hucksters of the information age' (Winner 1986) are understandably keen to concentrate on beneficial aspects of technologically-mediated life, there is a very real (and some would argue an inevitable) potential for ICTs to exacerbate some existing societal divisions and inequalities. If new technologies are to fulfil their potential to reshape societies positively as has been predicted, then governments need to carefully and systematically plan the ways in which they make use of ICTs. In trying to create a benign and beneficial role for ICT within the public sector, a host of problems, pressures and predicaments come to bear which require serious and prolonged attention. As the fraught and often flawed integration of ICT into UK government services has shown over the last five years, the information society will not be built in a day.

This need for political sensitivity towards the social aspects of ICT has long been acknowledged by some. From the 1970s onwards, authors such as Alvin and Heidi Toffler, Tom Stonier and John Naisbett produced best-selling depictions of technologically-led new eras. The beginning of the Thatcher/Reagan era in the West was awash with visions of prosperous and flexible information-led societies transformed by new technological 'waves' and 'megatrends'. The feeling of an imminent societal change was soon reinforced by a range of influential social forecasts from Europe and the USA, such as those of Masuda (1981) and Nora and Minc (1980). After a slight hiatus, the commercial rise of the internet during the 1990s then prompted policymakers and those in the public sector to sit up and take serious note of the implications of the information society. Once internet use began to move beyond the rarefied confines of the scientific and military communities and into domestic and work settings, it became increasingly clear to politicians and public alike that the information society represented more than a form of science fiction prophesy. It became a timely description for a fast-changing world.

Perhaps the most perceptive commentator at the time of these changes was the academic and government advisor Manuel Castells. In observing the development of what he termed the 'network society', Castells carefully spelt out the economic and social implications of the emerging informational society, moving the public and political debate out of the realms of futuristic speculation and into the sharp realities of global economics. For Castells, one of the key features of the information society was the 'networking logic of its basic structure' (1996: 21), brought about primarily by technological developments as well as the restructuring of capitalism and statism throughout the 1980s. As he argued:

> Dominant functions and processes in the information age are increasingly organised around networks. Networks constitute the new social morphology of our societies and the diffusion of networking logic substantially modifies the operation and outcomes in the processes of production, experience, power and culture. While the networking form of social organisation has existed in other times and spaces, the new information technology paradigm provides the basis for its pervasive expansion throughout the entire social structure.
>
> (Castells 1996: 469)

In Castells' eyes the significance of the network society and ICT in general is primarily one of global economics. Indeed, Castells and those who have followed him are explicit in their portrayal of contemporary patterns of economic and social activity as depending ultimately on the dynamics of the global economy, the 'network enterprise' of modern multinational corporations and the networking of labour in the form of 'flexi-workers'. In making these links, much of the network society thesis builds upon earlier writings on the post-industrial society. Here authors such as Alain Touraine (1969) and Daniel Bell (1973) mapped society's shift from an agrarian to an industrial basis, followed by a modern-day shift from an industrial to a post-industrial society. As in the later accounts of the network society, these authors acknowledged the importance of the service sector, the rise of professional, scientific and technical groups and, most importantly, information technology as all being the new 'axial principles' in the post-industrial society. Such a society was predicated on the commodification of information and growth of telecommunications technology. This, in turn, had fundamental implications for 'the way economic and social exchanges are conducted, the way knowledge is created and retrieved, and the character of work and occupations' (Bell 1973: 14).

The accounts of Bell, Touraine, Castells and others all highlight the key roles that education is seen to play in the fortunes of post-industrial economies. One of the most widely discussed areas of economic change within these literatures is the apparent restructuring of paid work and officially organised production and services. For example, new technologies such as the computer are accepted by most to have assisted the expansion of the services sector, and contributed to an associated shift in patterns of employment. ICTs are seen to have played no small part in the shift away from the traditional 'Monday to Friday, job for life' towards the individualisation of working patterns and the rise of 'flexi-workers' and careership. Specifically, the knowledge economy is seen to be predicated on quick-response, just-in-time production of goods and services, with countries, corporations and communities increasingly requiring workers and citizens with flexible, 'just-in-time' skills, competencies and knowledge. As Mossberger *et al.* observe, this is both an opportunity and a threat for workers and governments alike:

> A broad economic restructuring has widened economic disparities, automated some jobs out of existence, created new types of jobs, modified organisational practices, and altered traditional career ladders. In the 'new economy', workers are more likely to hold a number of jobs over a lifetime. Less-educated workers have watched their standard of living erode, and skills demands are increasing even in jobs requiring only a high school degree or less.
>
> (Mossberger *et al.* 2003: 61)

For many commentators, a ready solution to these challenges is to be found in education. Educational policies are now justified by notions such as '90 per cent of all new jobs require some ICT capacity' (Lewis 2002). Indeed, there is seen to be an increased 'informisation' of work – whether in the primary, secondary or tertiary sectors of production. Castells (1999) talks of 'information agriculture' and 'informational manufacturing', indicating that such industries are relying, more than ever, on using technology to process information and knowledge. The underlying message is twofold: (i) an information society and 'knowledge' economy requires an information-skilled workforce in order to succeed; and (ii) the key to an information-skilled workforce is education and learning.

The importance of learning and education in the information society – the rise of the 'le@rning society'

The prominence of education and learning within the post-industrial, information society analysis was in no small part responsible for the high-profile reassessment of education and training by educators and politicians in developed countries over the latter half of the 1990s. In the UK this was infamously embodied in New Labour's 1997 election commitment to concentrate on 'education, education, education'. The information society and knowledge economy agenda were particularly evident in the rise to political favour during the 1990s of the broad concept of 'lifelong learning', a notion embracing not only the compulsory phases of education but also education throughout adult life (Faure *et al.* 1972; Coffield 1997). Although it would be overstating the case to argue that many countries have since developed coherent strategies with respect to education and training through the life-course, the change in the political climate was unmistakable. In the UK, for example, official pronouncements about lifelong learning proliferated spectacularly from 1997 onwards; including three substantial reports (Kennedy 1997; Fryer 1997 and Dearing 1997), a major Green Paper (DfEE 1998) and a White Paper on the re-organisation of post-16 education and training (DfEE 1999).

This lifelong learning drive involved, and continues to involve, more than a narrow technical adjustment to the organisation of educational provision. It is an attempted transformation in learning opportunities in order to meet the implicit demands of the information society/knowledge economy. If, as previously discussed, it is accepted that the production and distribution of knowledge and information are increasingly significant processes in the determination of global economic competitiveness and development, which are reflected, in turn, in economic growth, employment change and levels of welfare, then the capacity of organisations and individuals to engage successfully in learning processes of a variety of kinds is an obvious determinant of economic performance (Pantzar 2001). For some commentators, the notion of the information society implies nothing less than a fundamental transition from an industrial to a knowledge-based *learning society*, with a need for education throughout the life-course predicated upon the unstoppable juggernaut of contemporary patterns of economic and social change (OECD 1996; Leadbetter 1999; Lundvall and Johnson 1994).

In this manner, a learning society has come to be seen as one 'in which all citizens acquire a high quality general education, appropriate vocational training and a job ... while continuing to participate in education and training throughout their lives' (Coffield 1994: 1). A successful learning society is therefore predicated upon a comprehensive post-school education and training system in which everyone has access to suitable opportunities for lifelong learning. Within this society, provision of education should be both excellent and fair leading to national economic prosperity and social integration. Although it remains a disputed notion, this is a fair summary of what the 'learning society' is deemed to be in British official discourse. As Quicke enthuses wryly:

> The idea of a 'learning society' certainly appeals as a vision of a 'better' world … it comes with all the connotations associated with learning and develop- ment. It conjures up images of a society which is open, democratic and forward looking, where citizens are provided with opportunities at any stage of life to acquire the knowledge and skills for self-development as well as for the benefit of others; and of organisations which are adaptable and innovative and which treat their employees as persons to be developed rather than resources to be exploited. Moreover, it is an idea which appears capable of generating a new consensus about the aims and purposes of education by breaking down the barrier between the 'world of work' and the 'world of education'. Many of the key goals of liberal democratic education – autonomy, respect for others, the development of critical capacities – are now seen to be more compatible with the kind of 'competencies' required by the polity and the economy. The scene appears to be set for the development of an education system which produces flexible pegs for a diversity of holes with ever-changing shapes!
> (Quicke 1997: 139–40)

The problem faced by politicians and policymakers is that the development of the learning society (defined in these terms) is by no means an easy task and entails manipulating complex economic and social processes. On the one hand, it holds the promise of increased productivity and an improved standard of living. On the other, it simultaneously implies that individuals and organisations face major challenges in adjusting to new circumstances. The emergent forms of economic activity affect the characteristic nature of work and the types and levels of skills required in the economy. As a result, the security and general quality of jobs are being radically altered, with profound implications for the welfare of individuals. Thus it is recognised that the nature of access to learning opportunities has implications not only for general economic competitiveness, but also for the employability of individuals and the consequent impacts on their standards of living. The dominant view has centred around the effective organisation of learning opportunities as being a crucial driver to both social cohesion and economic growth (Brown and Lauder 1996). The policy implications of this analytical approach are, of course, profound. Employability is based on the skills aspects of labour, and the enhancement of those skills is meant to be supported by lifelong learning from pre-school to post-retirement (Otterson 2004). Employees require not only good levels of general education, but also the capacity to adapt flexibly to changing skills requirements throughout their careers. Moreover, educational institutions should be organised in ways which ensure that these standards of general education are attained and also that the renewal of skills through continuing education and training is facilitated. As Frank Coffield (1999) has argued, however, one of the striking features of lifelong learning policies in countries such as the UK and US is that they have concentrated very much on the implications of this analysis for the formal education of individual learners.

Crucially this emphasis on individuals reflects a model of participation in lifelong learning which is based upon a simple notion of human capital theory where individuals participate in lifelong learning according to their calculation of the net economic benefits to be derived from education and training (Becker 1975). Given the dominant consensus about the general direction of economic change towards more knowledge-based forms of production, this logic sees workers seeking to participate in lifelong learning in order to capitalise upon the benefits which will flow from skills renewal and development. In this account, the principal issue which government policy is required to address is to ensure the removal of the impediments or 'barriers' which prevent people from participating in education and training. Achieving a learning society often thus comes to be defined in these ostensibly simple terms.

The importance of adult learning within the learning society model

Although the fortunes of the learning society are predicated on participation in education throughout the life-course (from 'womb to tomb' as a British trade union leader once indelicately phrased it), heightened attention has been directed of late towards what has traditionally been seen as the weakest sector of the lifelong learning process – *adult* education. Despite its new-found political prominence, adult education remains a rather disjointed and often ill-defined sector of education. Beneath the umbrella terms of 'adult education' and 'adult learning' lie a range of different forms of teaching and learning – reflecting the fact that adults can learn for a variety of reasons, in a variety of ways and in a variety of settings. For example, adult learning may be undertaken for personal or professional reasons (although the reasoning often blurs as time passes). Adult learning can be *formal* (institutionally sponsored and structured), *non-formal* (non-credentialised but still institutionally-based and structured) or occur *informally* by chance or during everyday activities. 'Adult education' in its broadest sense can take place in the workplace, college, home and community, at different times for different purposes.

Formal adult education is perhaps the easiest to identify and by far the most discussed form of adult learning in the academic literature. A multitude of institutionally provided and credentialised learning opportunities exists for adults above and beyond those provided by continuous post-compulsory education. In the UK, as in educational systems around the world, adult education institutions offer a range of full-time and part-time opportunities on a face-to-face or distance basis. Work-based training and education also represent major sources of adult formal learning – including health and safety training, work-related evening classes as well as basic induction training. *Non-formal education*, on the other hand, is a more ill-defined but no less important element of adult learning – referring to all organised learning which does not lead to formal qualifications but is nevertheless formally structured and provided. This notion is often erroneously used in official definitions where 'informal learning' is seen as non-course-based learning that leads on to more structured, planned learning (e.g. Cullen 2000). In this sense the third category of *informal learning* is more specific than simply being anything outside formal education but its importance in the learning society model is rarely discussed. As Tough (1978) contended, informal learning is the submerged bulk of the iceberg of adult learning both in terms of its visibility and significance. As such it is worthwhile taking the time to develop a clear understanding of informal learning, given its potential bearing for our later analysis of ICT-based learning.

At one level informal learning can be seen to simply be learning 'which we undertake individually or collectively, on our own without externally imposed criteria or the presence of an institutionally authorised instructor' (Livingstone 2000: 493). This non-institutional and unstructured definition is expanded upon by the European Commission:

> Informal learning is learning resulting from daily life activities, related to work, family or leisure. It is not structured (in terms of learning objectives, learning time or learning support) and typically does not lead to certification. Informal learning may be intentional but in most cases it is non-intentional (or 'incidental').
>
> (European Commisssion 2001: 33)

Whereas formal learning is typically institutionally sponsored, classroom based and structured, informal learning 'is not typically classroom based or highly structured, and control of learning rests primarily in the hands of the learner' (Marsick and Watkins 1990: 12). The most common form of informal learning is work-based 'learning on the job' – 'the common transformation of novices into highly proficient practitioners by their experiences of practice' (Hager 2000: 281). Indeed, as much as 90 per cent of new learning in the workplace has been estimated to be acquired through informal learning activities (Lovin 1992; Lohman 2000). Yet informal learning also includes a range of learning stimulated by general interests outside the workplace. These may include household work-related informal learning (such as home maintenance and renovations, cooking, gardening), community volunteer work-related informal learning as well as other general interest informal learning (Livingstone 2000).

The significance of ICT and adult learning in the learning society and information society models

It is our contention that the 'information society', 'knowledge economy' and 'lifelong learning' all coincide in, and are encapsulated by, the specific case of ICT-based adult learning. As we shall now go on to discuss, there are a number of compelling reasons why ICTs may be capable of making the learning society a reality – a set of arguments we shall refer to throughout the book (perhaps rather clumsily) as constituting the 'le@rning society' thesis. In other words, the 'le@rning society' refers to the practical achievement of the learning society vision through technological means. There are many influential and well-respected commentators who are convinced that ICT-based adult education is a ready means (perhaps the *only* means) of achieving the flexible, full participation in education required by the twenty-first century knowledge economy. In setting out to carefully examine the provenience for these claims we must first review the arguments *for* the ICT-based adult learning revolution.

It is certainly not hard to detect enthusiasm for ICT-based adult learning within the educational literature, reflecting the proliferation of new technologies such as the computer and the internet in adult educational settings. In essence ICTs are argued to make learning more effective and more equitable – to offer a diverse range of learning opportunities to a diverse range of adult learners on a convenient and cost-effective basis. Technology has been heralded by some to facilitate learning which is 'eclectic, holistic and flexible' (Friesen and Anderson 2004: 682). In short, ICTs are portrayed as making real the wider goals of the knowledge economy and information society:

> [We] stress the need to adapt European education and training systems both to the demands of the knowledge society and to the need for an improved level and quality of employment … In particular, Member States should strengthen their effort towards the use of information and communication technology for learning.
>
> (Council of the European Union 2002: para. 9)

Unpacking the specific rationales underpinning this acknowledged need for ICT in adult education can often be a confusing task, not least because there are many different types of ICT-based learning under discussion. Writers will happily eulogise about computer assisted learning, computer aided instruction, virtual education or 'e-learning' without ever specifically defining what they are referring to. This lack of clarity of definition was reflected in the UK government's working definition of 'e-learning' as *any* form of learning 'with the help of' information and communications technology tools (DfES 2002a: 2). Similarly, a recent EU action plan claims that 'e-learning refers to the use of new multimedia technologies and the internet to improve the quality of learning by facilitating access to resources and services as well as remote exchanges and collaboration' (Debande 2004: 191).

The plan talks of the 'awe-inspiring power of the internet to transform the educational experience … of the information age' (p. 191), and provides four distinct rationales for the spread of ICT. These are the social advantages (education reflects the concern of society over the spread of ICT, and seeks to demystify it), the vocational (the human capital model), pedagogical (the transformation of education from information transmission to genuine learning interaction), and catalytic (facilitating the transmission of skills for disadvantaged communities).

Thus, it is important for us at this early point in our investigation to recognise that ICT-based adult learning spans a continuum from ICT-based distance learning, which is exclusively provided and accessed on a remote basis via the internet, all the way through to the occasional use of technology in a predominantly 'face-to-face' classroom setting. In this sense there are a range of ways in which new technologies and learning intersect – from using ICTs as delivery mechanisms of information through to using ICTs as means of creating knowledge and providing tailored assessment and learner feedback. All of these could, and should, be seen as ICT-based learning.

This catholic definition is argued by some writers to be an impediment to rigorous questioning of the ICT-based learning society. As Curtain (2002: 5) argues, such diversity of provision and application 'make it impossible to offer a conclusive answer in most cases' to broad questions of the 'effectiveness' of ICT-based adult education. That said, there are perceived advantages of ICT-based learning in its many forms which are worth examining in preparation for our later empirical exploration of the le@rning society. In particular we will now briefly consider three main arguments for the integration of ICT into adult learning:

- that ICT can lead to a widening of educational participation;
- that ICT can support a diversity of educational provision;
- that ICT can lead to 'better' forms and outcomes of adult learning.

ICT and the widening of participation in education

We start with the common argument that ICT-based learning is a dynamic means of providing education for those adults who previously have not participated in formal or informal learning. It therefore acts as 'a catalyst for educational diversity, freedom to learn and equality of opportunity' (Forman *et al.* 2002: 76). For some, this potential alone makes the ability to learn with and through ICTs the key to establishing full participation in education throughout the life-course, and therefore making real the rhetoric of the le@rning society:

> E-learning is a relatively new tool with the potential to *radically improve participation* and achievement rates in education … Through e-learning we have the opportunity to provide *universal access* to high quality, relevant training and education.
>
> (DfES 2002b: 4, emphasis added)

In particular, ICTs are seen as opening-up access to learning across all groups which have historically been found to be under-represented in the adult learning population, such as the unemployed, the disabled, mothers, carers, the busy or simply the disinclined. For example, much attention has been directed recently towards the appropriateness of ICT for 'freeing' education from the barriers that may prevent older adults from participating. Specific barriers to learning for older adults, such as financial cost, the difficulty of physically travelling to sites of educational provision, the need to learn at an individualised and/or slower pace, and lower levels of previous contact with educational institutions, are now seen to be resolvable through the use of technology. ICTs such as the internet are argued to deliver educational opportunities to older adults on an easy basis, facilitating contact with communities of similar learners regardless of proximity, and generally making learning more convenient and flexible. Older adults, therefore, have the chance to become empowered 'silver surfers' where 'IT and the internet has the power to transform their lives ... 24 hours a day, seven days a week through the click of a button' (Ian McCartney in Cabinet Office 2000).

For all disadvantaged groups – from the elderly to the unemployed – the underlying argument is that distributing education opportunities via ICT (either fully or as part of 'face-to-face' provision) helps overcome the barriers which deter adults from existing kinds of formal learning. ICTs can make learning provision more flexible, bring costs down, make learning more accessible and affordable, offer reliable and accessible information, and allow people to learn at their own pace (e.g. SEDL 1995; Benton Foundation 1996; Glennan and Melmed 1996). In particular, ICT-based learning is seen as offering potential learners the fundamental advantage of *convenience* in terms of time, cost, family and work commitments (Pérez Cereijo *et al.* 2002; Learning and Skills Council 2002). Barriers to learning, whether they are situational (to do with lifestyle), institutional (related to the opportunities available), or dispositional (personal knowledge and motivation), are now seen as resolvable through the use of technology which can offer education to learners on an 'anyplace, anypace' basis. As Hawkey concludes:

> Rather than sitting in the stands or cheering from the touchline, ICT will enable learners to acquire transferable skills and to play a full part in the game, according to their own rules ... ICT can provide for learning that is differentiated by learner choice, rather than by the imposition of the governing body or the expert referee.
>
> (Hawkey 2002: 5)

ICT leading to a diversity of education provision

As well as overcoming entrenched barriers on the part of the individual (non)learner, ICTs are popularly seen as being able to support a diverse range of sources of educational provision from which learners can choose. In this way, ICTs are an ideal means of complementing and extending the wide range of 'traditional' providers of adult education such as firms, commercial organisations, further education institutions, adult education centres, charities and community education groups. In particular, by being able to negate practical issues of economy and scale (such as buildings, staffing and other physical resourcing limitations) 'virtual' educational provision is argued to offer providers a 'level playing field' – which particularly benefits smaller and more specialist organisations in distributing learning opportunities. ICTs are also argued to allow greater collaboration between private and public providers, fostering 'remote' partnerships between different agents as well as supporting the existence of wholly 'new' educational providers. In this way neo-liberal commentators have been especially vocal in welcoming ICTs as enabling a more competitive (and therefore more effective) 'marketplace' for adult education to develop.

Indeed, ICTs have long been seen as ready vehicles for the transformation of adult education along market-driven lines. Osborne and Gaebler (1992), for example, predicted an adult education market built around 'smart' credit cards, electronic information kiosks and a readily accessible computer database measuring the performance of adult education providers around the country. As well as supporting an increased diversity of formal educational provision, ICTs are also seen as increasing the opportunities for non-formal and informal learning to take place. In terms of taking learning into people's homes, communities and work-places, ICTs are seen to provide new 'spaces' for learning, and to blur the distinction between formal and informal episodes. As Dede speculated over twenty years ago, the biggest educational potential of ICTs such as the computer may well be that of initiating and facilitating informal learning outside the classroom:

> One important outcome of using the new educational technologies will be that a larger proportion of the society will have access to instruction. This expanded clientele for education will be interested primarily in informal activities; the learning experience will not necessarily involve credential-isation, nor will formal schooling be necessary. A variety of people, in their homes or in places of employment, will be able to interact with some device capable of improving the quality of their work, personal interactions or leisure activities. Barriers of inconvenience (in schedule or distance), cost and total time commitment will be reduced; the diversity of possible offerings will be expanded.
>
> (Dede 1981: 243)

Twenty-five years after Dede's comments, digital technologies are now seen to be successfully continuing the lineage of 'older' mass media and communication technologies (such as newspapers, printed books, public libraries, radio and television) which have contributed to the massification of self-education since the industrial revolution. Now technologies such as the computer and internet are seen to be intensifying individuals' access to educational opportunities, supple-menting and often superseding the spoken word and written page with a 'screen-based' context for self-education. Central to this is the role that ICTs can play in the reorganisation of adult learning – providing individuals with an increased accessibility and diversity of information and therefore freeing up the barriers and 'activat[ing] processes of self education' (Shuklina 2001: 73).

ICT as leading to 'better' forms and outcomes of learning

Tied in with this apparent 'democratisation' of adult learning is a third argument that ICT-use stimulates different and, it is contested, 'better' forms of adult learning – in particular the types of transformative learning which are argued to be 'the ultimate goal of adult educators' (Pascual Leone 1998). For many educational technologists, ICTs are welcomed as supporting and encouraging 'constructivist' learning. This refers to the use of ICT beyond an 'instructivist' mode of using ICT merely to acquire knowledge and skills, but involving more interactive and self-directed uses of ICTs which encourage learners actively to construct new ideas, concepts and meaning, therefore transforming their existing knowledge (Rosen 1998). In particular, ICTs are seen to fit in with constructivism's emphasis on learning mediated by participation in a social process of knowledge construction directly linked with the use of social artefacts or 'tools' to provide scaffolding for learning (Blanton et al. 1998). As Lim (2002: 413) explains, 'the emphasis is on the individual learning with a wide variety of tools and people that help them carry out their goal-orientated activities in a sociocultural setting'. This has led many educators to endorse the potential role of ICTs in supporting and mediating constructivist learning based upon personal control and the kind of diverse interactions that it is possible for learners to achieve with 'distributed knowledge' (Bostock 1998). As Crook (1998: 380) concludes, ICTs can provide ideal environments to support the constructivist perspective on learning, 'complementing and extending what is [already] powerful in existing practice'.

Some researchers have suggested, for example, that there are no great differences in academic performance between adult learners using ICTs (be it to either partially or fully learn) and their 'traditional' counterparts across a variety of different populations of learners and settings – from postgraduate classrooms to adults with mild learning disabilities (e.g. Schwartz and Duvall 2000; Anderson and Nicol 2000; Maki et al. 2000; Tolmie and Boyle 2000). Others have contended that ICTs can lead to 'better' outcomes of adult education. For example, it is popularly argued that ICT-based education and training in its diverse forms more often than not leads to improved learning attainment. Although, as with most research on education and technology the evidence base for the effectiveness of ICT-based adult learning is fragmented, proponents of this argument point towards a growing body of case studies which purport to demonstrate the 'value-added' of different forms of ICT-based adult learning. For example, a host of studies across the USA, Europe and Australia claim to provide evidence of improved academic performance and attainment via ICT. Curtain's (2002: 6) examination of different forms of online learning in Australia found that students' satisfaction with online learning which involved high levels of interactivity was 'on a par with the student satisfaction levels for classroom-based courses'. Other studies have also suggested that more interactive forms of learning with ICT can lead to more reflective, 'deeper' forms of learning and more empowered and democratic discussion amongst adult learners (Doubler *et al.* 2003; Jeris 2002) as well as proving to be an attractive and motivating medium of learning for adults with basic skills (Lewis and Delcourt 1998).

Mapping the adult le@rning landscape

The growth of ICT in education practice

From just this brief review it should be of little surprise that ICT-based learning has been enthusiastically seized upon by many in the adult education community. Although most practitioners and commentators would acknowledge that the implementation of ICTs is not without the usual operational difficulties and 'techno-hassle', many are keen to harness the potential advantages spelt out above. This has been reflected in an exponential growth in ICT-based adult learning provision *on offer* in the UK during the past ten years. The 'technologification' of adult learning has perhaps been most noticeable in the formal provision of ICT-based learning in adult education institutions, further education colleges and the like. In terms of resourcing and appearance, the physical makeup of the average adult education classroom has changed dramatically to reflect the rise of learning through new technology. Visitors to an adult education institution stand a good chance of being met by banks of sleek, flat-screened personal computers in computer labs, electronic whiteboards in classrooms and internet 'hot-spots' and cybercafés. In terms of ICT-related provision there has been a spectacular boom in the provision of learning *about* new technology in adult education institutions over the past ten years (Walker 2004).

The impact of ICT on the provision of adult education is also reflected in the growth of profit and not-for-profit organisations now offering different types of ICT-based adult education. Across the world, large-scale national learning organisations using ICT as their primary means of provision have been developed, such as the UK 'University for Industry', the Korean 'Cyber University' and the Spanish 'National Distance University'. Private providers of 'e-learning' such as the multi-billion dollar Skillsoft, Digitalthink and Pathlore corporations have established themselves as substantial training providers, as have innovative public/private sector hybrid organisations such as the UK National Health Service's 'corporate university' (reputed to be the largest e-learning project of its type in the world). Wholly private provision also seems to be thriving. Although estimates vary, the marketplace in 2005 for work-related 'e-learning' in the UK was estimated to be worth between £2 billon and £3 billion (Paton 2003). This is largely due to the plethora of private sector organisations now providing training programmes for millions of learners in the workplace. A multinational provider such as WebCT claims to have over 10 million licensed learners on their software systems. Skillsoft, the self-styled 'world's largest e-learning company', cites over 2,000 client companies and over 4.5 million registered online learners on a global basis.

Alongside the rise of the home as a key site of adults' access to and engagement with ICT-based learning, the community-based provision of ICT-based learning has also flourished. This has been most notable in the public provision of 'shared' or 'open' access to ICT in community sites such as libraries, museums, colleges, schools and purpose built sites. These are designed to complement and extend the longstanding public provision of ICTs by other organisations such as commercial 'shop-front' facilities and community organisations (see Liff *et al.* 2002; Casino *et al.* 2002; Todd and Tedd 2000). In offering 'safe and accessible environments' in which individuals can access new technologies, public ICT sites are argued to provide 'the human face of the information society' for many people who would otherwise be excluded (Stewart 1999: 1). As with the increase in work-based provision of ICT training, this community access is seen as leading to increased inclusion across a range of social and economic spheres, including the 'construct[ion of] self-educating citizens' (Hand 2005, in press). Public ICT sites are therefore argued by some commentators to play a powerful role in allowing peripheral localities a means to 'catch-up' with urban areas and to reduce regional 'information gaps' (Borgida *et al.* 2002). There are suggestions that such patterns of usage are leading to the inclusion of 'minority' groups of ICT users, with users of public sites appearing to be evenly distributed in terms of age and gender (Todd and Tedd 2000). For example, Liff *et al.* (2002: 88) reported the public ICT sites in their study attracting 'a socially inclusive mix of users' in terms of gender and age when compared to patterns of home ICT use. A range of other studies have also highlighted the varied social mix of people who use commercial and community internet cafés for learning (Uotinen 2003; Wakeford 2003; Lægran and Stewart 2003).

The growth of ICT in education policymaking

This growth of provision has been supported by governments through a range of associated policies and initiatives. Indeed, it would be fair to say that technology-based adult education now forms one of the predominant 'global policyscapes' across industrialised nations (Ball 1999). For example, in the United States, as in nations across Europe and East Asia, the role of technology in post-compulsory education has been of key political concern for the last ten years. The 'Digital Divide to Digital Opportunity' proposal introduced under the Clinton–Gore administration included the continued financing of Community Technology Centres and Neighbourhood Learning Centres in 'low-income' urban and rural communities, as well as the provision of computer access and training to adults and school children in these communities. Although some of these funding streams were cut under the subsequent Bush administration, finances were still being committed at the time of writing to promote tele-education and to fund the rural ICT infrastructure. Similar initiatives have been introduced across Europe, South America, East Asia and Australasia. From the German 'IT in Education: Communication rather than Isolation' programme to the Indian 'IT for all by 2008' initiative, governments in most developing countries (and many developed ones) have firmly stated their faith in ICT to establish inclusive learning societies. These initiatives, coupled with the ever growing rates of domestic and work-based access to computers, digital television and mobile phones, are now prompting politicians and educationalists to make wide-ranging claims about the combination of adult education and new technology as finally overcoming existing social inequalities and leading to a revitalisation of lifelong learning.

Of all the developed nations, the UK can perhaps consider itself to be the leading exponent of ICT adult education policymaking. Since its election in 1997 the New Labour administration under Tony Blair established a firm commitment to state-sponsored public provision of ICT and ICT-based education within its wide-ranging 'information age' policy agenda (e.g. Central Office of Information 1998). A raft of ICT-related policies focused around various strands of 'modernising' the organisation of not only education and learning but public services in general with the UK government devoting £7.4 billion to 'e-government' projects by 2006 (Caulkin 2004). These policies have centred on establishing the electronic delivery of public services to citizens, addressing social inequalities in the use of ICTs and improving the UK's economic competitiveness through the up-skilling of the workforce. From the early stages of this policy drive, public ICT provision was seen to underpin the inclusiveness and effectiveness of all these objectives. As Tambini (2000: 11) contended, unless all citizens were quickly provided with access to the technology required to make use of these ICT-based public services then any government's efforts would 'look increasingly illegitimate, as citizens that have paid for those services will have no access to them'. Thus, at the turn of the century the UK government set the ambitious target of achieving '*universal*' access to the internet by 2005, and *all* citizen transactions with the government being able to take place online by 2008 (Cabinet Office 2000). In this way, state commitment to 'universal service' was actually a commitment to a community-based universal service built upon the notion of 'post office' provision, so that citizens in urban areas have a public ICT facility within one mile of their home and those in rural areas have one within five miles (Pinder 2001).

Unlike other elements of the information age agenda, New Labour's plans regarding public ICT provision were relatively quickly (and relatively painlessly) put into practice. Starting in 1998 over 7,000 ICT centres were established in public sites throughout the UK, with the 'People's Network' initiative also establishing public internet connections in England's 4,300 libraries alongside a host of smaller-scale initiatives aiming to bring ICT and ICT-based services to those without. These English efforts were replicated by the Digital Scotland and Cymru Arlein initiatives developed by the devolved Scottish and Welsh governments. The majority of centres offering public ICT provision were established in existing sites such as schools, colleges of further education, libraries and museums, although concurrent efforts were made to utilise 'new' social contexts such as football stadia and pubs. The focus of all these initiatives on community-based technological accessibility was reinforced in New Labour's second term of office by a series of financial announcements concerned with extending levels of home access to ICT among the UK population. For example, the 'Computers Within Reach' initiative offered target populations access to low-cost re-conditioned computers.

All of these public provision policies were fashioned around the common purpose of providing increased opportunity of access to the material, cognitive and social resources needed to engage with ICT (van den Berg and van Winden 2002). Initiatives such as UK Online not only aimed to widen engagement with ICTs and assist the development of ICT skills, but also had wider remits to reduce current inequalities in participation in education and the labour market as well as increasing levels of citizen engagement with public services. Indeed, the underlying focus of much of New Labour's information age agenda during its first two terms of office went beyond simply facilitating the use of ICT but also aimed explicitly to increase individual levels of education and 'employability' – often through the means of computer skills provision and other forms of 'e-learning'.

In particular, the UK government pursued a parallel set of policy initiatives aimed at a technological re-engineering of the adult education sector under the aegis of the 'University for Industry' (UfI) and 'UK Online' initiatives. UfI was designed most prominently to take the form of a telephone-based helpline and website for directing individuals to approved and kite-marked learning opportunities as well as providing its own technology-mediated learning opportunities via a network of 'learndirect' centres in community sites throughout the UK. The initiative not only aimed to widen participation and achieve a 'mass-market penetration of learning' (Limb 2003), but to reduce the inequalities in participation amongst those groups traditionally under-represented in adult education. In 2004 the UfI initiative claimed that in its first six years of operation they had supported over 2 million educational enrolments for over 1.5 million online and classroom-based learners – nearly two-thirds of whom had not participated in formal learning for at least three years. This makes learndirect the self-proclaimed 'largest government-backed e-learning network in the world' (Crace 2004: 15) and suggests that if any country could lay claim to have established itself as a le@rning society then the UK would be a prime candidate.

Conclusion

In this chapter we have attempted, briefly, to set out the theoretical case *for* ICT-based adult learning. At a 'macro' level we have traced the prescient economic and societal pressures coming to bear on adult learning in developed countries as the concerns of the knowledge economy and information society become ever more tangible. On a more practical level, we have suggested some compelling reasons why ICTs have been so readily appropriated by those in education as a means of engaging individuals in the education and training required throughout their lifetime to survive, and thrive, as workers and citizens in contemporary society. In this way the range of policies put into place by governments to encourage and stimulate engagement with ICT and create 'learning societies' are based on a diversity of intentions and expectations. In theory, as we have seen, ICTs have the capability to *widen* educational participation to all individuals (especially those not previously taking part). In theory, ICTs can be used to support a diversity of forms of education – both formal and informal – and lead to better, more stimulating and more effective forms of adult learning. Although most educationalists would stop short of proclaiming ICT as a silver bullet for the woes of adult education, few would argue against the premise of the multi-billion dollar initiatives such as Ufl, UK Online and the like.

It is here that most academic discussions of ICT and adult learning would stop. Many in the usually contrary, sceptical and questioning academic community have proved uncharacteristically reluctant to challenge the 'orthodoxy of optimism' which has grown up around the notion of the le@rning society. Most books on ICT and learning do little more than elaborate on the arguments set out in this chapter – highlighting the potential of ICT to transform education and sometimes going on to present case studies of the application of technology in various learning scenarios. Whilst such an approach is all well and good, our intention in this book is to test empirically the claims outlined in this chapter. To what extent is the le@rning society thesis a robust and accurate description of the UK in the early twenty-first century? To what extent is it a likely prediction of the UK a decade from now? In testing these claims we also aim to examine empirically a set of less-widely espoused arguments which challenge the le@rning society thesis. These more controversial arguments (often encompassed under the 'neo-luddite' or 'dystopian' banners) take the stance that ICTs actually do at least as much harm as good in realising the aims of lifelong learning. It is to these counter-arguments we now turn.

Chapter 2

Impediments to adult learning in the digital age

Introduction

While the le@rning society thesis is a compelling one, even its most enthusiastic supporter would concede that many of the claims upon which it is founded remain untested. As a field of academic and practitioner endeavour, educational technology has promised much over the past 30 years but could be criticised for delivering rather less. Notwithstanding the undoubted *potential* benefits of new technology for adult learners, it is important for educationalists and policymakers to restrain from unconditionally assuming ICT-based education and training to be a universal panacea for educational problems. For all the perceived benefits of ICT there are a set of corresponding caveats, drawbacks and unresolved problems which tend to be ignored or summarily dismissed by some in the educational technology community. This chapter balances the claims and contentions made in Chapter 1 about the educational promise of ICT by outlining some of the potential limitations of ICT-based adult education. These can be seen both in terms of technological and social impediments to the realisation of the le@rning society vision.

Technological caveats to the le@rning society

A substantial stumbling-block to the realisation of the le@rning society thesis lies in the ambiguity of ICTs when it comes to matters of equality. Although new technologies may have the potential to overcome barriers to education their use is also just as capable of introducing new forms of impediment to full participation in education. In other words the use of ICTs could compound rather than alleviate the problem of widening participation in lifelong learning. First, the ideal of technology-based learning for all runs contrary to the current marked inequalities in people's access to and use of ICTs. Although the magnitude of figures vary, the enduring trend is that even within technologically developed regions such as North America, western Europe and south-east Asia, specific social groups are significantly less likely to have ready access to ICT (e.g. NTIA 1995, 1999, 2000; Bonfadelli 2002; Loges and Jung 2001; UCLA 2000; Jung *et al.* 2001; Dickinson and Sciadas 1999; Reddick 2000; Lenhart *et al.* 2001; Quibria *et al.* 2002; Rainie and Bell 2004; Bromley 2004). Differences are apparent along nearly every 'social fault line' which characterises social exclusion in general – i.e. age, income, socio-economic status, ethnicity, geography and gender. The so-called 'digital divide' is a marked feature of any current learning society, and there is strong evidence that these inequalities are strengthening over time rather than being reduced. At one level, as the US Department of Commerce has outlined, these divisions are simple and stark:

> [some individuals] have the most powerful computers, the best telephone service and fastest internet service, as well as a wealth of content and training relevant to their lives … Another group of people don't have access to the newest and best computers, the most reliable telephone service or the fastest or most convenient internet services. The difference between these two groups is … the Digital Divide.
>
> (US Department of Commerce 2000)

In some respects, education has a new underclass, with a gap in basic technology skills among adults greater than in literacy or numeracy. Around 24 million people in the UK are unable to use the internet, for example, but there are only 7 million with poor reading or writing (Whittaker 2003). As we discussed in Chapter 1, governments have introduced a range of initiatives to provide full or 'universal' access to ICTs, most notably via the provision of technology in public sites. However any attempt to address the basic problem of democratic access of ICTs runs the risk of facing the subtle but significant digital divides which remain once an individual has been given opportunities for physical access to ICTs. For example, people do not simply either have (or not have) access to ICTs – in the same way that people do not simply have (or not have) access to transportation or healthcare. It is important, therefore, to distinguish between an individual's *effective* access over their *formal* access to ICT (Wilson 2000). Although in theory the formal provision of ICT facilities in a community site such as a library or college mean that all individuals living locally have physical access to that technology, such 'access' is meaningless unless people actually feel able to make use of such opportunities. And even then, the quality of that access is not the same as that provided by owning a computer in the home or using one in the workplace.

The digital divide is not the simple premise which some commentators may have us assume. Political and popular understanding of the digital divide has tended to be in strictly dichotomous terms – you either have 'access' to ICT or you do not, you are either 'connected' or 'not connected'. From this perspective, as we saw from the public access policies outlined towards the end of Chapter 1, the digital divide is easily defined and, as a result, is easily 'closed', 'bridged' and 'overcome' given a political will to provide for those 'without' (Edwards-Johnson 2000; Devine 2001). This has promoted some commentators to blithely predict a near future where 'only the homeless and the jobless will be webless' (Sutherland 2004: 7). Yet, there are some very powerful counter-arguments to the universal access argument and the premise that almost everyone will soon be able to gain access to education and learning through ICTs. In particular, four prominent areas of the digital divide debate are beginning to be reconsidered by social scientists which could have a direct bearing on ICT-based adult learning:

- what is meant by ICT;
- what is meant by 'access';
- what is the relationship between 'access to ICT' and 'use of ICT'; and
- how can we best consider the consequences of engagement with ICT.

Reconsidering what is meant by ICT

Although the umbrella term 'ICT' refers to a range of different, albeit rapidly converging technologies, there is a tendency when discussing the digital divide to use either too narrow a definition of ICT in terms of specific technologies or else too broad a definition in terms of ICT as a homogeneous concept. An example of this latter tendency was evident in the *Economist*'s (2001: 10) assertion that 'ICT is spreading faster than any other technology in the whole of human history [and] ... the poor are catching up'. Even when not treating ICT as a homogeneous concept many commentators have been extremely limited in their definition of terms – content to define ICT vaguely in terms of *computer* hardware or exclusively in terms of access to the internet (e.g. Norris 2001). However, we know that people's use of technology extends far beyond the realm of the computer through technologies such as digital television, mobile telephony and games consoles; all important but disparate elements of the contemporary techno-culture as well as potential conduits of learning (see Choi 2002; Katz and Aakhus 2002).

This plurality of technologies is complicated further when the content that is provided via ICTs is considered – the 'soft'ware rather than the 'hard'ware. In other words, the digital divide can also be seen in terms of the information, resources, applications and services that individuals are capable of accessing via new technologies. In this manner a focus on content rather than technological platform is perhaps a more accurate and useful point of reference for the digital divide debate. World wide web resources, for example, are accessible through a variety of platforms – from computers and digital television to wireless telecommunications devices. It is clear that beneath the umbrella term of ICT we are concerned with a heterogeneous range of technologies and 'contents' – not all necessarily analogous to each other. Crucially then in terms of the potential establishment of the le@rning society, 'digital divides' should be seen as running separately through all of these technologies and forms of content.

Reconsidering what is meant by 'access'

These points lead us to a second area of contention – what is meant by 'access'. As it stands in contemporary debate 'access' is an often used but woefully ill-defined term in relation to technology and information (see Rifkin 2000). In policy terms 'access' tends to refer to making ICTs available to all citizens – in other words 'access' is used solely to refer to the provision of physical artefacts. This was apparent in the UK government's model of 'post office' provision outlined in Chapter 1. Yet this notion of access in terms of whether technology is 'available' or not obscures more subtle disparities in the *context* of ICT access. For example, accessing an online educational course from a home-based computer or digital television set is not equally comparable to accessing the same materials via an open-access work station in a public library or other community centre. Issues of time, cost, quality of the technology and the environment in which it is used, as well as more 'qualitative' concerns of privacy and 'ease of use' are all crucial mediating factors in people's 'access' to ICT.

As already mentioned, it is important to acknowledge the importance of an individual's 'perceived' (or effective) access in practice over the theoretical (or formal) access to ICT. Indeed, any realistic notion of access to ICT must be defined from the individual's perspective. Although in theory the formal provision of ICT facilities in community sites means that all individuals living locally have physical access to technology, this 'access' is meaningless unless people actually feel able to make use of such opportunities. The logic of this argument can be seen in the increasing numbers of public payphones in UK towns and cities that have been converted to offer email facilities alongside conventional telephony. Despite this formal provision it would be a nonsense to claim that every individual in these towns and cities now has effective and meaningful access to email or, indeed, equitable access to email when compared to individuals who use email from their home or place of work.

Instead of either having or not having access to these many different technologies in many different contexts it follows that access to ICT and the digital divide are hierarchical rather than dichotomous concepts. Indeed, as Toulouse (1997) observed, there are two distinct types of access: whether people have access at all and the hierarchy of access amongst those that do. Beyond the simple issue of 'access/no access' to ICT come more complex questions of levels of connectivity in terms of the capability and distribution of the access concerned. On a practical level for example, access to a personal computer does not guarantee a connection to the internet, any more than access to the internet is a guarantee of effectively accessing every available website and online resource. These issues have led some authors to refer to an 'access rainbow' of physical devices, software tools, content, services, social infrastructure and governance (Clement and Shade 2000) or 'various shades' of marginality between 'core' access, 'peripheral' access and non-access (Wilhelm 2000). As we can already see, the technological prerequisites to establishing the le@rning society are complex and not simply matters of physical access or purchasing power.

Reconsidering the relationship between 'access to ICT' and 'use of ICT'

It is also important not to conflate 'access to ICT' with 'use of ICT' and to assume that access to ICT inevitably leads to use. A widely held view amongst technologists is that ICT-related inequalities are primarily due to the 'S-curve' of expansion of technology use in society from groups of 'early adopters' through to the majority of the population at a later date. From this perspective there are distinct phases of the diffusion of innovations – an almost inevitable progression through 'innovators', 'early adopters', 'early majority', 'late majority' and then 'laggards' in terms of individual citizens (Rogers 1995). The logic of this 'natural' diffusion thesis is centred around the notion that widespread inequalities in the use of ICT are only a passing phase of technological adoption and that, in the long term, the only people not using new technologies will be 'information *want nots*' – refusniks who for ideological reasons chose not to engage with ICT despite being able to in practice. From this perspective the digital divide is merely a temporary stage of societal adoption of ICTs, as Tuomi infers:

> If we study available evidence, the digital divide is closing rapidly. During the last decade millions of people have gained access to computers every year. Never in the human history have there been so many people with access to computers, digital networks and electronic communication technologies.
> (Tuomi 2000)

But this assumes a monolithic view of technology, whereas rapid technological obsolescence means that the have-nots are always playing catch-up, and the S-curve is a permanent feature rather than a transitional phase. The danger of the determinist dismissal of the long-term significance of social inequalities is that it ignores the complex relationship between access to ICT and use of ICT. As just asserted, we should recognise that access to ICT does not denote use of ICT. Even more importantly 'use of ICT' does not necessarily entail 'meaningful' use of ICT or what could be termed as 'engagement' where the individual exerts a degree of control and choice over the technology and its content thus leading to a meaning, significance and utility for the individual concerned (Silverstone 1996; Bonfadelli 2002). Having made these distinctions we should see that an individual's lack of meaningful use of different technologies once having gained suitable conditions of access to them is not necessarily due to technological factors (such as a lack of physical access, skill or operational ability) or even psychological factors (such as a 'reticence' or anxiety of using technology) as is sometimes claimed by technologists. Instead, as a range of studies have shown, individuals' engagement with ICTs is based around a complex mixture of social, psychological, economic and, above all, pragmatic reasons. Technological engagement is therefore less concerned with issues of access and ownership but more about how people develop relationships with ICTs and are capable of making use of the social resources which make access useable (Jung *et al.* 2001; Garnham 1997). As Heller (1987: 20) argues, at best, technology offers a number of options, or 'choices based on particular contingencies', which determine the variable impact of technology on people. Thus, individuals' interactions with ICTs are not as simple as the 'user'/ 'non-user' dichotomy constructed by much of the previous literature and certainly not determined solely by issues of physical access to technology.

Reconsidering the consequences of engagement with ICT

In attempting a deeper understanding of the potential technological impediments to the le@rning society thesis we should also consider the fundamental yet often unvoiced element of the digital divide debate – the outcome, impact and consequences of accessing and using ICT. Indeed, much contemporary debate concentrates only on the means rather than the ends of engagement of ICT use. As Wise acknowledges:

> the problem with questions of access is that they reify whatever it is that we are to have access to as something central to our lives without which we would be destitute. They, therefore, redirect debate away from the technologies or services themselves.
>
> (Wise 1997: 143)

To be of lasting significance any consideration of ICT-based adult learning must combine questions of access and use of technology with the impact and consequences of engagement with information and communications technology for individuals. In this way, we examine to what extent (and why) the consequences of using and engaging with ICTs are not automatic for all. For example, we know that by its very nature some information is specialist and restricted to a few with the requisite intellectual and managerial skills to manipulate and use it (Lyon 1995). Therefore, the effects of accessing information, resources and services via ICTs are not uniform for all users. As Balnaves and Caputi reason, it follows that where the impact, meaning and consequences of ICT use are limited for individuals then we cannot expect sustained levels of engagement:

> The concept of the information age, predicated upon technology and the media, deals with the transformation of society. However, without improve-ments in quality of life there would seem to be little point in adopting online multimedia services.
>
> (Balnaves and Caputi 1997: 92)

In this sense, the consequences of meaningfully engaging with ICT could be seen in terms of the effect on individuals' 'social quality' – i.e. socio-economic security, social inclusion, social cohesion and empowerment (e.g. Berman and

Phillips 2001). Perhaps the most useful framework to utilise here involves the various dimensions of participation in society that can be seen as constituting 'inclusion' (e.g. Berghman 1995; Oppenheim 1998; Walker 1997). These can be grouped as: *production activity* (engaging in an economically or socially valued activity, such as paid work, education/training and looking after a family); *political activity* (engaging in some collective effort to improve or protect the social and physical environment); *social activity* (engaging in significant social interaction with family or friends and identifying with a cultural group or community); *consumption activity* (being able to consume at least a minimum level of the services and goods which are considered normal for the society); and *savings activity* (accumulating savings, pensions entitlements or owning property). The impact of ICTs could be seen in these terms which reflect the extent to which technology use enables individuals to participate and be part of society, i.e. the extent to which 'ICTs enhance our abilities to fulfil active roles in society, or being without them constitute[s] a barrier to that end' (Haddon 2000: 389).

Social and educational caveats to the le@rning society

Entrenched patterns of non-participation

Alongside these 'technological' caveats are a host of social and educational rejoinders to the arguments laid out in Chapter 1. Even if every adult has sufficient opportunity and means to make use of ICT for learning, the predictions outlined in Chapter 1 may still face considerable impediments to being realised. First and foremost is the issue of people's deep-rooted reluctance and/or inability to take part in any form of adult learning – whether technology-based or not. Indeed, inequalities in participation in all forms of education have endured over the past 50 years with significant minorities routinely excluded (e.g. Beinart and Smith 1998; NCES 2002a). Indeed, some of the most recent figures suggest that these patterns of exclusion are growing:

> Fewer than a fifth of adults say they are doing some sort of learning, the lowest figure since before Labour took office in 1997 ... Social class appears to be a significant factor in the trend ... the decline in participation has been marked among people from the poorest backgrounds.
>
> (Kingston 2004a: 15)

As well as the socio-economically disadvantaged we know that those currently 'disenfranchised' from formal and non-formal adult education and training (the non-learners) also continue to be more frequently female, not employed, older, less qualified, with lower literacy skills or with negative attitudes to institutional learning. Recent work has emphasised the role of regional influences in deter-mining patterns of participation (Gorard and Rees 2002). We also know that those employed in high-skilled, white-collar occupations are more likely to be recipients of work-based training – as are those in larger and multinational firms (OECD 2003). There are suggestions that participation levels in informal adult learning are similarly delineated. Crucially, those who currently do *not* participate can make up over one-third of adult populations in developed countries (NIACE 2003; OVAE 2000). These inequalities are entrenched and perpetuated by the fact that those who participate and benefit most from adult learning tend to be those who have higher educational attainment levels and continue learning throughout their lifetime (OECD 2003). The current growth in the UK of cycles of low-pay, low-status jobs, often created by attracting inward investment of a mercurial nature, does not fit well with the employability and vocational lifelong learning models described above (Jackson 2003).

These enduring inequalities of participation point towards persistent barriers that potential learners face which may not be as easily overcome by the advocates of ICT as is sometimes assumed. To a large extent they are presaged by the determinants of adult learning. A genuine widening of adult participation in learning must target those groups least likely to come forward, according to a recent review of the evidence (Macleod 2003). Those least likely to come forward have then been found to be also the most likely to drop out of a course even if successfully enrolled (Walker *et al.* 2004). Perhaps the most obvious obstacle that most people face when envisaging episodes of learning is the potential cost (Maguire *et al.* 1993). Whereas these costs can be of the direct kind (such as fees) they are more commonly indirect, such as the costs of transport, child-care, forgone income, time and even the emotional cost for those with families (Hand *et al.* 1994). These costs are clearly more restrictive for the poor, and to some extent for women, who are still faced with the greater burden of child-care, for which support is generally poor (NIACE 1994), and other domestic responsibilities (Park 1994).

The loss of time, particularly for a social life, is another cost of learning in some cases, especially in a country such as the UK with some of the longest average working weeks in Europe (McGivney 1993). Adult education is now suffering not so much from lack of leisure time but from the multiplicity of opportunities available for that time (Kelly 1992). Taking a course often involves an adjustment in lifestyle which may be possible for an individual, but is more of a problem for those with dependants or in long-term relationships. Relationships can be strained, particularly for women taking courses to progress beyond the educational level of a male partner. Women who are more likely to be employed part-time, less likely to be aware of opportunities stemming from the workplace and have domestic and child-care responsibilities, and generally poorer transport facilities, therefore face many threats to participation.

The institutional barriers to training can come from the procedures of the educational providers, in terms of advertisement, entry procedures, timing and scale of provision, and general lack of flexibility. People often want to fit learning around other tasks of equal importance in their lives, since they cannot always get time off (Park 1994). Drop-out from formal adult learning is commonly caused by people discovering that they are on the wrong course (Pyke 1996), with nearly half of students in one survey feeling they had made a mistake (FEU 1993), and part of the blame for this must lie with the institutions in not giving appropriate initial guidance. Many learners are disappointed by the lack of help available in choosing a course and in staying on it. There is in general a low level of awareness of sources of information and financial incentives for training.

Part of the cause of lack of training must be the lack of appropriate provision (Banks *et al.* 1992). Even those studying may not have found what they actually wanted. This is particularly true of non work-related training and learning for leisure (NIACE 1994), and is reinforced by the emphasis on certified courses, heavily backed up by the incentives in the funding arrangements to provide accreditation of all adult education. Taster courses of the kind leading to no qualifications are still being axed, due to a shortage of funding, and this is especially impacting on courses designed to remedy shortages of the basic reading and writing skills that make further, perhaps certificated, study possible (Hook 2004a).

Not only does this deny some people the opportunity to learn new interests and make new friends, it denies returners an easy entry route back into education, especially for women who may be more intimidated by examination demands, according to one study (Burstall 1996). Even where provision is available, knowledge of opportunities may be patchy for some parts of the population (Taylor and Spencer 1994), giving many a feeling that 'you are on your own'. In addition, an estimated 7 million adults in Britain may have difficulty with writing or numeracy, and perhaps one-fifth of adults have problems with basic literacy (Moser Report 1999). These deficiencies appear to pass through generations of the same families (DfEE 1996), reinforcing their importance as a 'reproductive' determinant of adult non-learning. Ironically, it is courses in basic skills such as literacy and numeracy that are rated among the worst offered by colleges (Hook 2003), with a severe shortage of appropriate tutors leading to a high drop-out rate among students.

Whatever barriers are faced, they are harder for the less motivated prospective learner, so the influence of lack of motivation to learn may be underestimated by literature concentrating on the more easily visible barriers such as cost and entry qualifications (McGivney 1993). Many people display an incorrigible reluctance to learn formally. In fact, it has been estimated that one in five adults form a hard core of non-participants outside all attempts to reach them (Titmus 1994). If all the barriers were removed for them, by the provision of free tuition and travel, they would still not want to learn. Lack of drive thus becomes the most important barrier of all, since it is seen as easier to get a job instead, and qualifications are seen as useless anyway (Taylor and Spencer 1994). Qualifications may even be seen as antagonistic to getting a job, and only concerned with entry to more education. Learning is something done early in life, as a preparation, but with no relevance to the world of adults (Harrison 1993).

There is a danger, therefore, that the ability of ICT to overcome all the above 'barriers' to education is over-emphasised by those seeking to establish ICT-based adult learning. As we have just seen, the barriers faced by non-learners are often considerably more than issues of convenience. Indeed, the chief obstacles to participation reported by adults are not necessarily the physical barriers of time and place which ICTs can address to some extent, but rather lack of interest and motivation (La Valle and Blake 2001; DEST 2001). Positively influencing individuals' decisions to learn may not be simply a case of making learning opportunities more 'convenient' via ICTs (Dhillon 2004). If adults have not previously engaged in learning and education due to issues of motivation and/or disposition then there is little reason to assume that ICTs alone will alter this. Whilst ICTs can overcome situational and institutional barriers they can perhaps do less to alter the social complexities of people's lives and the 'fit' of education in these lives. As Kennedy-Wallace (2002: 49) reminds us, 'whether learning online in the workplace, in college or at home, e-learning is still about learning and culture, not just technology and infrastructure'.

ICT may not lead to 'better forms' of learning

It can also be argued that different forms of technology-based educational provision are more conducive to certain types of learning over others – thus challenging the ability of ICT to effectively widen participation in *all* forms of learning. It has been argued by some educationalists that much current ICT-based learning is not the same as 'real-life' learning, appearing to be more about knowledge dissemination than a genuinely transformative process. In this way, serious questions have been raised regarding the pedagogical 'fit' of ICT – especially given the narrow paradigms favoured by many current providers of technology-based learning which tend to rely on one-way transmission of information and communication (see Hamilton *et al.* 2002). As we discussed in Chapter 1, ICT-based adult learning should be critical and emancipatory rather than solely about the transfer of information and determinate skills. Yet it is the latter which dominates much current online provision (Selwyn and Gorard 2002). As Mayes argues, for reasons of cost alone the interactive pedagogical opportunities offered by information and communications technology are often overlooked by adult learning providers, leaving ICT-based pedagogy rooted in more 'old-fashioned' linear and restricted models:

> There are really two pedagogies associated with ICT. One is the delivery of information – this is predominantly the pedagogy of the lecture or book, and emphasises the 'IT' – the other is based on the tutorial dialogue and involves conversations between tutors and students, and mainly emphasises the 'C'. Of course, successful teaching is underpinned by both – and the rapid interplay of the two is ideal – but in the context of lifelong learning policy the real problem is that 'IT' is cost-effective and the 'C' is not. Unfortunately, in terms of pedagogic effectiveness the second is better than the first.
>
> (Mayes 2000: 3)

Even the most enthusiastic proponents of ICT-based distance learning recognise the limitations of the medium for delivering all types of learning. As De Kerckhove (1997) concedes, at best ICT enhances rather than replaces 'real-life' learning. ICT should not necessarily be seen as providing *better* educational contexts, but *different* contexts for learning. For example, it is argued that in many learning situations reliance on the 'virtual' rather than the 'real' ignores the uniqueness of educational processes that are fundamentally altered once digitised and delivered online. This is especially the case with creative, ethical, moral and aesthetic learning (Trow 1999).

These limitations extend to the perceived ability of ICTs to facilitate 'trans-formatory' learning. As Imel (2001: 1) argues, 'one of the myths of ICTs ... is that they promote constructivist learning'. As Imel and others point out, although constructivist learning may be desirable, it is not dependent on the use of ICTs or even necessarily encouraged by them (Wessel 2000; Wilson and Lowry 2000). As with all types of learning, the effectiveness of constructivist learning via ICTs is not a certainty but very much dependent on the social contexts in which they are used. Studies which have looked at the use of computers for constructivist learning in classroom situations have found that they can just as easily result in 'ineffective' learning – merely reproducing the existing beliefs and practices of teachers and learners rather than automatically leading to 'better' forms of learning (Daley *et al.* 2001; Dirkx and Taylor 2001).

There are also many practical 'human' problems in learning with new technologies. This point has been highlighted by recent research which has examined the lack of human and social contact of many ICT-based learners. Learners through the internet, for example, can often feel isolated when presented with the unstructured nature of the data and its sheer quantity. Students in Hara and Kling's (2002) ethnographic study of online graduate education, for example, complained of the 'isolation' and 'ambiguity' of being 'left' to study some elements of their courses via ICT when compared to their 'offline' educational experiences. Other studies have also reported that learners feel lost and disconnected from the institutional systems within which they are learning when studying via ICTs (e.g. Sarojni *et al.* 2002; Persell 2004). As Connolly *et al.* (2001) report, although the need for self-tuition when learning via ICT may be motivating for some adult learners, others find the experience more problematic – citing, in particular, the lack of external human direction. This could be the biggest barrier to widespread ICT-based learning for, as Doring (1999: 8) observes, 'education is a fundamentally conversational business'.

ICT encouraging a limited provision of adult learning

In spite of the increasing amount of education being supported and provided through new technology there are concerns that ICT may actually be contributing to a narrowing of adult education provision – especially around business and industry-friendly 'core skills' and 'key competencies'. For example, it is noticeable that much current virtual and classroom-based provision concentrates on work-related skills and, in particular, ICT skills (NCES 2002b). Indeed, one of the criticisms of adult education at present is that there is a confusing plethora of ICT and 'computer literacy' qualifications (E-Skills NTO 2001; Devins *et al.* 2002). Whilst it is rational for commercial organisations to concentrate on the more profitable aspects of adult education and training, the notion of ICTs promoting a *broad* spectrum of learning opportunities may not, therefore, be fulfilled. If anything, current ICT-based educational provision could reinforce the priorities of 'traditional' post-compulsory educational provision on 'employability' and work-related skills at the expense of facilitating different forms of learning and competencies amongst different social groups.

If we examine the rhetorical basis for many governments' current educational agendas it soon becomes clear that ICT-based adult learning (and indeed adult education as a whole) is seen primarily from the economic perspective of up-skilling (or at least re-skilling) the workforce. As the director of ICT for the UK University for Industry remarked, 'we are there to make a difference to individuals to make them more employable' (Sutton 2003). In this instance, any notion of using ICTs to widen participation to all social groups could more accurately be seen as increasing participation amongst economically active groups. This narrowness of provision and emphasis on profitability is compounded by the relative cost-ineffectiveness of using ICTs for educational organisations. Indeed, as the OECD recently recognised, with ICT-based provision the marginal cost of a student remains close to the average cost. Thus, a powerful argument,

> never addressed head-on by the advocates of computer based learning, is that in traditional teaching or training based on the idea that a group of people benefit from the knowledge of a teacher or instructor, it often costs very little to add an extra learner ... [institutional] cost may disqualify e-learning as a panacea for adult learning.
>
> (OECD 2003: 190)

Clashes between 'old' and 'new' forms of adult learning

Alongside these organisational issues are concerns whether ICT-based education is actually complementing existing 'traditional' provision of adult education and leading to a diversity of provision. Indeed, the notion of ICTs leading to a greater co-operation and collaboration between different education providers remains unrealised in many countries, with ICT-based learning often seen as a threat by, rather than a complement to, existing education. The few studies which have looked at the implementation of ICT-based education from an institutional perspective point towards significant institutional clashes which can occur when technology-based 'solutions' are introduced into existing educational markets, especially where commercial and private interests are seen as impinging on the public good of education. This was evident in our own study of the implementation of 'virtual college' initiatives into the Welsh adult learning sector and the resultant 'surf and turf' wars (Selwyn and Gorard 2002). Contrary to the rhetoric of diversity, existing educational institutions were found to perceive new forms of ICT-based provision such as the UfI initiative as depriving them of learners who otherwise would have enrolled on traditional courses. It would seem that such institutional clashes are especially likely when 'new' ICT-based initiatives are positioned around pre-existing structures of education (see also Cornford and Pollock 2003). As Alger (2001) points out, collaboration between educational institutions is not automatic, with established providers keen to retain their 'traditional' reputations. There are also a host of legal issues such as intellectual copyright over shared learning resources, sensibilities over branding and corporate affiliations. It is perhaps naive for ICT-based education providers to believe that they can complement existing forms of adult education without being seen as a competitive 'threat' by the very institutions that they paradoxically rely on for partnership. The concern that large e-learning schemes may 'dilute the market share' of existing providers persists (Crace 2004: 15). One of the very real barriers faced by ICT-based education therefore is the pre-existing micro-politics that characterise all education sectors.

It follows that a crucial but often overlooked caveat to the le@rning society model is the integral role of business and industry and, in particular, the perennial clash between education both as a private interest and a public good. This problem is not necessarily a new one, as Tasker and Packham warned:

> The two worlds of [education and industry] remain profoundly different. The purpose of industry is to generate profit for private gain, usually in competition with other companies. The profit so generated may or may not benefit society; the concept of public good is not central to industry's concerns. The purpose of education is to generate knowledge through collaboration between scholars, not competition, and in such a way that society as a whole benefits.
>
> (Tasker and Packham 1993: 134)

Doubts should therefore be raised over the attractiveness of ICT-based adult learning programmes to private companies, especially if they do not prove to be immediately beneficial. Moreover, the financial incentive for companies to actively encourage participation among more socially diverse, but less profitable, consumers is even less obvious. As Robertson (1998) asks, if technology-based lifelong learning for all is that attractive a proposition then why had the private sector not already autonomously developed it without the need for encouragement and coercion from governments?

Avoiding assumptions of ICT as a technical fix for adult education

These points should raise sufficient doubts to persuade the objective observer to reconsider the inevitability of the le@rning society vision. Much of what is assumed about ICT and adult learning by its proponents stems from a viewpoint where virtually all of society's problems, be they economic, political, social or ethical, are subject to the 'technical fix' of ICT (Volti 1992). Of course, adult learning is not the only field where this reasoning occurs. Societal trust in the technological fix has been well established, in fields as diverse as warfare, medicine and environment (Weinburg 1966). In its current form, as Neil Postman (1993) observed, the overriding 'message' of the current wave of new technologies is that the most serious problems that confront society require technical solutions through fast access to information that is otherwise unavailable.

As we have seen in this chapter, there are enough caveats to the le@rning society thesis to steer us away from the technical fix approach alone, as a solution for adult learning. Yet in education, as elsewhere, questioning the value and worth of ICT has been generally frowned upon. Most commentators when approaching the role of ICT in education ultimately conform to one of the two dominant paradigms that beset discussions about technology – either technological or social determinism. On the one hand many political and academic commentators see ICTs as an inevitable consequence of technological development and change, adhering to an overtly *technological determinist* view of technology and society. As Woolgar (1996: 88) details, technological determinism is 'the belief that new technology emerges as an extrapolation of previous technologies, with the characteristics of a technology hav[ing] a direct impact on social arrangements'. This view sees technology as guiding and shaping society with its own logic, as an influence autonomous to social forces. Thus, technology is seen as a driving force of society. Where technology changes so society follows. Although it is difficult to find anyone actively aligning themselves with technological determinism today such discourse has been central in post-war thinking about technology in the Western world from Herbert Marcuse to Marshall McLuhan. As can be seen in Chapter 1, forms of technological determinism are prominent in much post-industrial writing about the 'information society', with authors such as Bell and Castells often attacked for this aspect of their thinking (Webster 2002).

A similarly pervasive societal discourse is that of ICT as more or less a neutral means to an economic, political or social end. This opposite view of the determinist relationship between society and technology asserts that it is society that shapes and influences technology. Thus *social determinism* posits that technology is a neutral instrument that can be moulded and used for various purposes. In this approach, technologies merely appear in response to society's demands and the interests of the market. As Wise (1997) asserts, from such a social determinist view technology is contingent on interpretation and interpretative frameworks. The properties of objects are not inherent in the objects themselves but conferred on objects by social consensus and definition. Thus, society shapes technology (as opposed to the technological deterministic view of technology shaping society). This view is inherent in the commonplace presentation of educational technologies being 'just a tool'. For example, ICTs are often positioned by proponents of the le@rning society as appearing in response to the educational needs and demands of adult learners. Currently, there is much reporting of e-learning sweeping through the workplace in response to employee and employer demand for 'just-in-time' training. Adult education and training can now have what it is 'keen to embrace'. Yet one can argue that the social determinist treatment of ICT is equally as constricting as the technological determinist discourse. As Bromley asserts, to view ICT in purely instrumentalist terms is to overestimate its flexibility and neutrality:

> Recognising the significance of the context of use, the responsiveness of technologies to social dynamics, is a useful insight, but technologies are not infinitely malleable; they cannot be put to absolutely any end at will, and certainly not with equal ease.
>
> (Bromley 1997: 54)

Thus, as Winner (1980) argues, although social determinism acts as an antidote to 'naïve' technological determinism, it is itself flawed in assuming that the technologies themselves are of little importance. In short, social determinism fails to recognise that technological artefacts are imbued with politics in their own right. To view new technologies as neutral tools arising from societal demand may be to misread the social and cultural significance of the technology itself.

Conclusion

This tradition of 'misreading' technology and society within the existing educational literature should act as a warning as we now move on to pursue our own empirical analysis. There is clearly a need to step beyond the limitations of previous analyses if we are to develop deeper understandings of ICT and adult learning. We need to be aware of the social, cultural, political, economic and technological aspects of ICTs – the 'soft' as well as the 'hard' concerns. Echoing Qvortrup's (1984: 7) argument that such questions 'cannot be properly understood if we persist in treating technology and society as two independent entities', this book strives to move beyond the view that technology is distinct from society in either its cause or effect. Instead we make a conscious effort to move away from positions of either 'technophilia' or 'techno-neutralism' towards a perspective that avoids drawing a clear technology/society distinction, but is aware of the social contexts where technologies and policies are developed, and focuses on the ones where they are used. Such an approach is not intended to be wilfully obtrusive, but an attempt to test the assumptions underlying the current turn to adult learning in the digital age (not to say the billion dollar initiatives based upon it) from a more objective social-scientific perspective.

We have seen in this chapter that there are a variety of reasons why ICTs may have little, if any, beneficial impact on existing patterns of lifelong learning – in terms of establishing a learning society with full and flexible participation in education through the life-course. We have also discussed the theoretical need to approach ICT and adult learning as inherently entwined with the social – not acting as autonomous forces on the lives of individuals but bound up with the everyday social, political, economic and cultural nuances of day-to-day life. It must be remembered, however, that the caveats posited at the end of Chapter 1 equally apply to the arguments presented in the present chapter. Many of the suggested limitations outlined in this chapter are as speculative, conceptual or predictive as the arguments presented in Chapter 1 – with little empirical foundation or testing. There is therefore a danger of rejecting the le@rning society thesis on equally subjective grounds as some may rush to accept it. In avoiding a utopian, overly determinist view of technology and learning we must be wary of instead 'falling into a more current trap – an orthodoxy of pessimism where nothing good can be said about information technology' (Wresch 2004: 71). Thus we need to develop a rigorous, objective framework for empirically testing the cases *for* and the cases *against* the le@rning society model. How this can be achieved is now discussed in Chapter 3.

Questioning adult learning in the digital age

Research questions and methods

Introduction

Throughout the previous two chapters we have made the case for researchers to move beyond a view where ICTs may have 'universal' effects (for better or worse) on adult education. Instead we have suggested that research should start from the premise that the implementation of technology in adult education is intrinsically entwined with the social, economic, political and cultural aspects of adult education. We therefore start our own empirical investigation of adult learning and technology from Fitzpatrick's observation that:

> New technologies do not emerge *ex nihilo*, but are always embedded within social contexts whose contours shape the ways in which technologies are constructed and utilised.
>
> (Fitzpatrick 2003: 133)

This approach has fundamental implications for how we set about researching adult learning and technology. For example, with regard to the key theme of ICT widening levels of educational participation we must recognise from the outset that (non)participation in learning or use of ICT facilities is often entwined with wider exclusion from society – and not merely due to a set of technical 'barriers' which can be addressed directly. Increased access to ICTs has been found not to lead to greater 'universal' empowerment in areas such as medicine and health, social security or political engagement (e.g. Henman and Adler 2003; Nettleton and Burrows 2003) – so imagining education to be unique in this respect is to ignore the evidence of wider impacts of ICT on society.

The problem with attempting to take account of these disparate shaping factors is that it can quickly make researching adult education and technology seem fruitless. Taking a more socially sophisticated approach to researching the le@rning society is unlikely to provide *definitive* answers to the questions most often asked by politicians and funding agencies. Adult learning, as with education in general, is a socially complex and often contradictory activity which makes providing 'one-size-fits-all' conclusions and sweeping judgements unwise, if not impossible. Although many in education are currently clamouring for conveniently packaged 'cause and effect' explanations, we cannot expect to 'prove' that ICT leads to 'improved' or more 'effective' adult learning. We are unable, for example, to conduct a randomised trial of the appropriate length and scale. That said, we believe that our research approach results in findings and recommendations which will be of lasting long-term value. By carefully positioning our research questions and methods of inquiry within the wider social realities of adult learning and technology we hope to get 'under the skin' of a much discussed but little examined area of education. Our overall aim is therefore a simple one – to provide a more realistic and accurate account of how and why ICTs are actually being used and not used by adult learners.

Avoiding some weaknesses of previous research on adult learning and technology

Many studies of adult learning and technology choose to ignore these wider social contexts and, instead, concern themselves exclusively with the learner and the technology. This narrow focus is often deliberate – allowing researchers to concentrate on the important questions of how technology 'works' in terms of learning and cognition and ultimately leading to the improvement and refinement of ICT applications. One danger in studying ICT and adult learning along these 'research and development' lines lies in the temptation for researchers to 'go native' and lose any objective critical sense of what they are studying. For many researchers of education technology the fact that ICTs are of universal educational benefit is not a matter for debate. In this manner, studies of ICT and adult learning are sometimes compromised by a desire amongst practitioners, politicians, funding bodies and researchers to 'prove' once and for all that technology is having a positive and transformative impact on people's learning. Some researchers, for example, will discount negative findings or apparent shortcomings as short-term and unimportant – assuming that any deficiencies are likely to be transitional rather than of long-term significance. The research which has been carried out into technology and learning can therefore be criticised for often being rich in enthusiasm but lacking rigour (Maddux 1989; Selwyn 2000), with the seductive nature of technology distracting some commentators from the pressing need for objectively constructed and executed research. As Daniel Menchik observes:

> While some writers have conducted methodologically sound, empirical research, there are many … who have gained prominence on the basis of their anecdotes and platitudes appropriating 'transformative' or 'empowering' characteristics to cyber-education.
>
> (Menchik 2004: 202)

The relatively uncritical body of existing research has left many of the assumptions outlined in Chapters 1 and 2 untested and, in turn, left many educationalists content to assume that ICTs are relatively unproblematic in their ability to reduce barriers to learning. To date there has been little rigorous empirical evidence pointing either way. From an academic point of view this lack of empirical evidence severely compromises our theoretical understanding of participation in education and of the roles that ICT may play (see also Dillon 2004; Wellman 2004). From a practitioner perspective, adding to our empirical understanding of participation (and non-participation) is crucial to using ICT to promote genuine inclusion if this is at all possible (Sargant 2000). Although there are large bodies of both support and scepticism, we still know little of the extent to which access to home and community-based ICT is contributing to learning amongst adults in the UK. In particular, very few large scale analyses have been carried out examining the success (or otherwise) of recent ICT-based educational initiatives as well as the impact on adult learning of the proliferation of ICTs into people's homes and workplaces. Similar criticisms can be levelled against our empirical understanding of general ICT use and non-use throughout the adult population. It would seem that the information society, knowledge economy and lifelong learning have been the subjects of much speculation but little careful examination.

Principles of the Adult Learning@Home research project

With these issues in mind, we set out to develop a rigorous framework of investigation that could begin to examine the nature and impact of ICT across adult learning in its many forms. In developing our research design we were keen to pursue a course of enquiry that paid attention to the quality, objectivity and representativeness of the data which were collected and therefore resulted in findings which were capable of producing as realistic an understanding of technology and adult learning as possible, whilst remaining generalisable to a wider population. Based on our reading of the existing literature (as set out in Chapters 1 and 2) we developed a set of broad guiding intentions for our own research.

Researching the present, rather than the potential, uses of technology for adult learning

As we have seen, many authors have produced detailed accounts of what technology-based adult learning could look like given the necessary time and resources. Not so many authors have concentrated on what ICT use, and what adult learning, is actually taking place at the moment – for better or worse. For many authors the rationale for adopting (what could be politely termed) a 'forward thinking' perspective is largely pragmatic. Given that educational technology is a perpetual 'work in progress' which is always developing and changing, there is a danger that any attempt to examine present applications would be rendered obsolete after only a short period of time. Additionally some technologists would argue that new technologies such as the internet are only in their infancy and so are not ready for meaningful critical examination.

Whilst these points should certainly be borne in mind by researchers they do not, in our opinion, hold sufficient weight to preclude a meaningful examination of the present-day application of ICTs and ICT-based adult learning. A rapid pace of change is apparent with *any* element of social life or social phenomenon. Such is the rate of change in schools and colleges, for example, that any classroom-based research project is almost inevitably outdated once it is completed – a fact compounded by the painfully slow publication and dissemination process used by academics. Neither are new technologies always as nascent or fast-moving as some commentators would have us believe. Educational computing, for example, has been in fairly common use in schools, colleges and universities for 40 or so years. Home computers have been a mainstream consumer item for nearly 30 years. To claim that much of what these technologies are used for in the home, workplace or classroom has changed radically during this time is to underestimate the conservative nature of many people's computer use. The lynchpin applications for most computer users continue to be word-processing, spreadsheets, databases and sending text messages between networked computers – all applications with a history of 30 years or more and, for the last decade, mostly implemented on an IBM-compatible PC running a variant of Windows. Even 'new' policy initiatives such as the Ufl or learndirect have now been in existence for seven or eight years. Whilst technologies and policies will obviously continue to develop and alter they can be considered sufficiently embedded within the adult learning landscape to bear up to scrutiny and empirical examination.

Researching all types of adult learning

Our second guideline is a reaction to the tendency for education researchers, quite understandably, to focus on formal learning taking place in educational institutions. Again, the reasons for this practice are rooted in pragmatism as well as reflecting the composition of the educational research community. With many researchers still actively involved in formal education where their main experiences and interests lie, there is an understandable tendency to focus research there. Informal learning has also been traditionally avoided by educational researchers for a number of other pragmatic reasons, not least those of adequately defining and measuring it (Girod 1990). Moreover, informal learning (however it is defined) is also more difficult to 'capture', more expensive to research and more prone to disruption from research. As a result informal learning has been considered by academics as a 'less serious' and certainly less viable focus of research than formal learning (Edwards and Usher 1998).

Yet given the potential of ICT in supporting and stimulating informal learning – as well as the apparent prevalence of such types of learning within the learning society – it seemed counter-intuitive for us to design a study of adult learning and technology which only focused on the use of ICT for formal learning in formal educational institutions. Thus, as we shall go on to describe, within the Adult Learning@Home project we actively sought to widen the perspective of our research to consider learning and technology use of all types and in all contexts.

Gaining a representative picture of engagement and non-engagement

A third guiding principle was to avoid a tendency for educational researchers to over-privilege those more 'visible' individuals who are making regular and sustained use of ICT and/or are known learners. The temptation for researchers to go 'looking for learning' in the field is a strong one. Asking questions of those we know to be engaged in learning – and preferably engaged in ICT-based learning – is less time-consuming and expensive as well as more likely to yield relevant data in terms of our research questions. But while we can learn much about the potential of ICT-based learning by deliberately focusing our attention on known ICT-based learners, adopting such an approach will do little to address some of the fundamental issues which emerged during Chapters 1 and 2. Questions about the inclusiveness of ICT-based adult learning, for example, can only be answered by an inclusive research process which asks questions of everybody – be they lifelong learners or non-participants, ICT users or not.

This desire for representativeness within the research process led us to pay particular attention to those individuals who do *not* participate in adult learning as well as those who do *not* make use of different technologies. Of course developing a focus on non-learners is not a natural thing for educational researchers to do – with the result that those who appear not to learn tend to be largely invisible in the academic literature. However, in the case of ICT-based adult learning this is perhaps the most important section of the population to develop a detailed understanding of. Thus, we were aware that our research should reconsider why people engage and do not engage with adult education – ICT-based or not. In doing so we decided to take a deliberately non-pejorative approach by not assuming ICT use or participation in learning to be necessarily desirable, beneficial, empowering or 'normal' activities for all individuals. Similarly, we did not approach non-participation as some kind of 'deficit' on the part of the individual concerned. Instead we attempted to develop a more nuanced and sophisticated reconceptualisation of people's non-use of technology and non-participation in education. As will be seen in later chapters, this approach allowed us to reach some powerful findings and conclusions which otherwise may not have been apparent.

Taking a lifelong perspective of people's learning and technology use

Becoming a 'learner' or a 'computer user' is not a one-off transition or permanent transformation. For instance, having used the internet once does not necessarily make someone an internet user for life. This simple distinction is often poorly represented by researchers seeking to document inequalities of participation and engagement. Media and communication studies researchers, for example, are now beginning to document the 'phenomenon' of a minority of the computer-using population who report no longer using the internet despite having done so previously. These are being presented by some researchers as 'defectors' or 'drop-outs', prompting discussion of what may be causing this disengagement. Of course the reality of people's (non)engagement with technologies or education is much more fluid and nuanced than the pejorative labels of 'drop-out' or 'defector' suggest. Other social scientists (ourselves included) are instead beginning to consider the notions of people's engagement with learning and/or technology throughout their lifetime in terms of 'careers', 'pathways' or 'trajectories' – concepts which acknowledge that people's motivations, circumstances and actions are almost always in a state of evolution and, therefore, flux. Even the most dedicated learner or heavy user of technology will develop, refine and alter the quantity and quality of engagement as their life circumstances alter. Thus, in designing our research we aimed to avoid merely producing a 'snap-shot' account of people's learning and technology use at one period in time. Instead we attempted, where possible, to provide longitudinal accounts of how technology had played a role in people's lives and people's learning throughout their life-course. Of course, given the constraints of a 27-month research project, much (but not all) of this involved eliciting *retrospective* accounts from our respondents. Nevertheless we tried to adhere to the maxim that lifelong learning merits a lifelong approach to the questions asked and the data collected.

Developing an understanding of the different contexts of learning and ICT use

Adults' use of ICTs for learning can take place in many different social contexts other than the classroom. The title of our research project was an initial attempt to reflect this, although we quickly came to realise that ICT use and learning occurs in many contexts other than the home. As well as the classroom and the home, any attempt to attain a representative picture of people's technology use has to consider workplaces, libraries, community centres and, with the advent of increasingly powerful mobile technologies, high streets, pubs and even public transport. A crucial question to ask here is to what extent people's technology-based learning may be shaped by these different social contexts and, conversely, to what extent these social contexts are being shaped by people's technology-based learning. Taking the home as an example, in terms of the physical confines of the house, issues such as whether broadband cabling is in place and the available space in living rooms and bedrooms can have immediate influences on the ways technologies are used. Conversely, many people consciously or unconsciously shape their houses around technologies, as is often the case with the physical, spatial and aesthetic accommodation of home cinema systems in people's front rooms or the furnishing of a home office or study. Less apparent, but no less important are the regimes and relationships which characterise the home and household. For example, the informal 'rules' which build up within the household around computers or television sets are crucial to how technologies are used (for example, the negotiated understandings of who 'owns' the technology, who controls it, and what time is allotted for what uses).

Similar issues are at play in the workplace, community sites, and even with mobile technologies. Although it is possible to overplay the importance of the social contexts of technology use, we felt it important to gain a sense of the contexts within which ICTs were being used and not used, and not to view each social context in isolation but attempt to gain an understanding of how the different contexts interacted, complemented or clashed with each other in the lives of learners and non-learners. How individuals then negotiate these different tensions, and how this then influences (non)engagement with technology and adult learning, are critical questions for a study of technology and adult learning to address.

Taking a perspective rooted in the perspective and experiences of the individual

Finally, we took the approach from the outset of our research that the outcomes and effectiveness of ICT-based adult learning are only really apparent in the lived experiences of the individual adults who are the ultimate 'end-users' of the policy initiatives, technologies and education provision concerned. Our project and this book are therefore unashamedly based upon the voices of individual adults. It is surprising how much research on adult learning does not pay much attention to the adult learners themselves, leaving many accounts of ICT and learning crucially lacking an acknowledgement of individuals' agency in their (non)use of technology and (non)engagement with learning. Throughout our work we have instead endeavoured to examine ICT-based adult learning from an avowedly 'bottom-up' perspective. As Chatman (1996: 205) reasons, 'the process of understanding begins with research that *looks* at [an individual's] social environment and that *defines* information from *their* perspective' (emphasis in original).

Research questions

Without further labouring the point, there are many areas to be addressed in terms of developing a realistic and thorough understanding of adult education and ICT. These were encapsulated within the deceptively simple set of questions posed in the preface to this book which still need to be asked of the 'learning society' and 'information society' discourses. Put crudely, *who* is engaging with *what* technologies, *where*, *when* and *why*? A final overriding question is perhaps the hardest one to answer – *what* are the contexts and implications of this engagement? With all these points in mind we developed a set of research questions to encompass the technologies that people were using in their everyday lives, the activities which they were using them for, the skills and strategies involved in this use, as well as the meanings, motivations and contexts implicit in this engagement. With this in mind, the remainder of the book focuses on two principal areas of concern:

- In what ways does access to ICT in the home, workplace and other community settings contribute to learning amongst adults?
- To what extent does the use of ICT interrupt or reinforce existing patterns of participation in lifelong learning?

These concerns are encapsulated in five specific research questions:

- *What are the established patterns of lifelong learning that can be documented amongst particular adult populations?* From our previous research and reading of the existing literature we already know that formal participation in lifelong learning is likely to be patterned in relation to a number of key variables, such as: previous educational experience and employment history, age, sex, geography and ethnicity. In order to document any possible impact of ICT on participation it will be necessary, as baseline data, to document established patterns of formal participation in lifelong learning amongst a systematic sample of the adult population in a range of carefully selected community contexts.
- *Who, amongst those populations, has access to what forms of ICT within the home, the workplace and wider community sites?* As discussed in Chapter 2, a prerequisite to learning through ICTs is having effective access. We need to document both who has formal access to what ICTs (through domestic ownership or proximity to work or community based facilities), and the hierarchy of access amongst them. We also explore how access is patterned according to individual factors such as age, sex, class, geography (both in terms of distance and terrain), ethnicity and previous education biography, as well as family and household dynamics.
- *What do adults within those populations use ICTs for and how does ICT use fit into their lives more generally?* Here we are concerned with why, and for what purposes, individuals are using (and not using) ICT. Thus, we are interested in exploring which individuals are choosing to use ICT for formal and informal learning opportunities, what learning provision they are opting to use (be it locally, nationally or internationally provided) and the motivations under-lying these choices. We are also interested in documenting the full range of apparently 'non-educational' use of ICTs – how new technologies such as computers, the internet and digital television fit into individuals' wider 'techno-cultures' and their lives as a whole, and how the use of such tech-nologies co-exists with activities not involving technology.
- *How do adults learn to use ICTs effectively for formal and informal learning activities?* Learning to use new technologies poses a range of learning challenges which are tightly related to the purpose of using the technology. We need to explore how different groups of adults approach learning to use ICTs such as computers for both formal and non-formal learning activities. In so doing we will explore the structuring resources that are used in supporting learning whether these are derived from home, from learning materials or elsewhere. We will also explore the conditions of learning associated with informal home-

based learning. Are the conditions of learning for informal use of the computer the same for adults at home and in community provision? What are the conditions of learning that are necessary for effective formal learning in such non-institutional settings?

* *What are adults actually learning through their engagement with ICT environments?* Evidence from previous research involving children suggests that engagement with ICTs outside formal educational settings can give rise to many different forms of learning only some of which are related directly to those sponsored within formal educational programmes (Facer *et al.* 2003). It is necessary to document the full range of different forms of learning experienced by adults as they work within different ICT environments on a range of different activities. This will involve, for example, examining the role of play and the role of different forms of collaborative activity with others both face-to-face and electronically.

Research methods

Carefully constructing a set of research questions is of little use unless equivalent care is also taken in developing the research design. Here, also, many previous studies into adult learning and educational technology can, in our opinion, be found lacking. Although educational research overall has been criticised of late for its lack of rigour, relevance and weak execution, research into adult learning is generally considered to be an especially weak sub-discipline. Although held in slightly higher regard, social research into technology use is also of varying quality – scathingly characterised by one commentator as consisting of little more than 'pundit suppositions, travellers' tales and laboratory studies' (Wellman 2001: 2031). As Mossberger *et al.* (2003: 16) conclude, the provenance for many research findings, such as those cited so far in this book, 'is indeed murky'.

With a need for rigour in mind, an initial concern was with achieving a 'goodness of fit' between our research questions and our methods of data collection. It is surprising how often this basic stage of the social research process is apparently flouted by researchers in the area of adult learning and technology, as can be seen in recent studies' use of woefully inappropriate research methods for the questions they set out to address. For example, attempting to measure the barriers to online learning via an internet-based survey would (or at least should) appear to be counter-intuitive – yet has been attempted recently and repeatedly by at least one set of researchers (Berge and Muilenburg 2004). The same can be said for the nature of the questions which are asked – another common weakness of some educational technology research. This is typified in the proud repeated findings of one group of US researchers that skills with computers and internet use predict whether or not students buy textbooks online (Yang and Lester 2000, 2002; Yang *et al.* 2003).

One of the particular weaknesses of the existing evidence base on adults, technology and learning is the general lack of large-scale quantitative data-sets which are generalisable beyond their immediate sample. As O'Neil (2002: 77) notes with regard to studies of community ICT-use, there is 'an increasing amount of qualitative data on these types of projects, but little research generating quantitative data'. Conversely, there have been calls for the need to collect fine-grained qualitative data. It has been argued that survey data alone cannot afford a deep understanding of adults' use of ICTs and, in particular, the social contexts in which ICTs are being used. As Haddon and Silverstone argue:

> Although standard quantitative measures of diffusion, possession, patterns of use *etc.* can be useful they do not tell us the phenomenological experience of these ICTs, their meaning, their role in the home or their significance. They do not tell about pleasures derived from consuming ICTs, nor about the anxieties or conflicts which they create. And they do not show us how people are constantly involved in attempts to keep control and regulate these technologies.
>
> (Haddon and Silverstone 1994: 2)

Similar arguments have been made in terms of the need to collect qualitative data when investigating adult learning in its many forms. Zborovskii and Shuklina (2001: 80) reason, for example, that quantitative data alone cannot capture adequate understandings of informal learning, which is 'by its very nature, a deeply personal and individual type of activity. In order to investigate it, therefore, subtler methods are needed that require involvement in the individual's life'. Given the scope of the research questions we posed at the beginning of the research project it was decided to adopt a large-scale, mixed-methods approach to our data collection. This was to be based around a 'tiered' approach with the findings from one method feeding directly into the design of the subsequent methods (Gorard 2004).

Mixed methods research has been identified by a number of authorities as a key element in the improvement of education research, perhaps because research claims have greater impact when based on a variety of methods (National Research Council 2002). It can also lead to less waste of potentially useful information, for if social phenomena have multiple empirical appearances, then using only one method in each study can lead to the unnecessary fragmentation of explanatory models (Faber and Scheper 2003). The problems for qualitative work conducted in isolation include those of generalisation and warrant. Simple quantitative work can supply the 'what' and the 'how many', while basic qualitative work can illustrate the 'how' and 'why'. For longitudinal or retrospective work concerning the life-course of individuals, it is recommended that we use quantitative data concerning the structure of the life-course, and qualitative data to interpret the experiences during the life-course (Erzberger and Prein 1997). This is largely what we do here via an approach we term 'new' political arithmetic (NPA, but also termed 'explanatory' two-phase research; Cresswell 2003).

Statistics as used in social science derive from the idea of political arithmetic in the 1960s (Porter 1986). Its purpose was to promote sound, well-informed state policy (applying therefore to the 'body politic' rather than the 'body natural' of the natural sciences), and its aims included raising life expectancy and population figures, and reforming health, education and crime. NPA is a development of this approach, adding a stage of using in-depth data to help explain the patterns in the body politic. In its simplest form it involves a two-stage research design. In the first stage, a problem (trend, pattern, or situation) is defined by a relatively large-scale analysis of relevant numeric data. In the second stage, this problem (trend, pattern, or situation) is examined in more depth using recognised 'qualitative' techniques with a sub-set of cases selected from the first stage.

In this way, researchers tend to avoid what Brown (1992) calls the 'Bartlett' effect of producing plausible but false results when basing an analysis solely on qualitative data, and they also avoid the simplistic answers often gained from numeric analysis alone. In NPA the explanatory phase collects new in-depth data, but in a focused attempt to elucidate the more general findings from the descriptive phase. Each type of data has a different purpose for which it may be best suited. The in-depth data in the second phase is coded for analysis using codes generated by the analysis of the more clearly generalisable data from the first phase. The first phase provides the patterning, and the second phase provides the explanation of those patterns. However, the model also allows the grounded coding of data for analysis in the second phase. In this way, the new ideas that emerge can then be used to re-analyse the data from the first phase. This is not so much a two-phase design as a fully iterative one.

In this project we used three stages of data collection: (i) a large-scale household survey; (ii) semi-structured in-depth interviews; and (iii) year-long 'ethnographic case-studies' of adult learners, their friends and family. These stages of data collection are briefly outlined in later sections of this chapter.

Choosing the research regions

One of our first considerations was where we should locate the research project. Although it is normal for a study of this size to select a diverse and well-spread sample, the region of this study is limited. This is so for three main reasons – the length of time under scrutiny, the need to uncover local determinants of participation and the need to identify local sites for public access to ICT (at least partly so that we could boost the sample of users of public-access sites). Our survey asked adults as old as 96 years about events as long ago as their birth. The study, therefore, encompasses nearly 100 years of life histories, and so to some extent the breadth of the study is curtailed by the depth necessary. Social changes, patterns of provision, and even the form of implementation of wider policies have local variations and so the role of place needs to be considered in the study, and it is only by limiting the study geographically that the necessary level of historical and background detail can be established. However, having identified the patterning in a local context it should then be possible to generalise the findings to a national context more safely than with many wider ranging samples since the merely local effects are now more clearly visible. In this way the study develops the sampling techniques pioneered by Gorard and Rees (2002).

Instead of focusing purposively on known sites of technological activity and 'best practice' we decided to research intensively a set of carefully chosen areas that could provide a broad and representative picture of England and Wales. Given the suggested educational, technological, economic and social importance of the idea of place to the project we felt it important that our research activities focused on a variety of localities within one region. Given the authors' then geographical location in Cardiff and the need to carry out in-depth 'face-to-face' work as well as a large-scale household survey, we first chose to focus the research on South Wales and the west of England. Studies carried out solely in Wales are sometimes dismissed by English-based researchers as somehow lacking in 'relevance' to the UK. Although we see this as an unfair criticism, our decision to focus on sites in both countries was intended to counter such prejudicial responses to our research findings as well as provide interesting cultural (and sometimes political) comparisons. According to official national statistics, the west of England is often considered to be a fairly 'technology-rich' area of the UK – whilst Wales (and therefore South Wales where the majority of the population resides) is consistently nearer the bottom of most 'league tables' of ICT access and activity.

The population area of South Wales and the west of England consists of twenty-one unitary authorities, twelve in Wales and nine in England. The area is made coherent by geographical proximity (often referred to as the 'M4 corridor') while at the same time including a variety of rural areas, such as Carmarthenshire and the Forest of Dean; the cities of Bristol, Cardiff, Swansea, Newport and Bath; the Welsh coalfield valleys; and a range of different sizes of town and rural village communities. Secondary data were used to characterise the authorities in terms of a range of social and educational measures capable of disaggregation at the ward level. These data provided the sampling frame for the survey and part of the context for the primary analysis. Since the study conceives patterns of participation in education and training as socio-economic trajectories over time, and uses primary data from at least as far back as 1910, both the current and the past characteristics of the study region need to be taken into account in selecting the representative study sites. While in some senses this makes selection of the sites for the study more complex, any sampling strategy cannot be valid for *all* of that period, and so if the stratification is less than perfect for current indicators it may not matter as much as for a typical cross-sectional survey.

First, a sample of four unitary authorities was drawn, and then three electoral wards were selected within each of those authorities. Selection of these areas was made on the criteria both of typicality within the population, and on the basis of theoretically interesting characteristics. In choosing unitary authorities, data relating to educational qualifications, rurality and poverty were examined, allowing us to categorise the authorities into four 'ideal' types (see Table 3.1 and Madden et al. 2002, for further information).

We chose two sites in England and two in Wales whilst including one site of each type within the final selection of four. Since the Welsh Valleys were only in Wales, and the typical small towns were only in England, there were only a few

Table 3.1 Four 'types' of unitary authorities

Type	Income deprivation	Population density	Educational deprivation	Number of authorities
Welsh valleys	High	Medium	High	9
City	Medium	High	Low	2
Smaller towns	Low	Medium	Low	5
Rural	Medium	Low	Medium	5

permutations to consider. For example, if we used the city of Cardiff as the city type, then the rural site had to be in England. If we used the city of Bristol as the city type then the rural site must have been in Wales. Since the rural areas were more typically located in England this argument favoured the choice of the city of Cardiff, Blaenau Gwent (the most deprived Welsh valley area), Bath and North East Somerset (the least deprived small town area) and the rural area of the Forest of Dean (with low population density and multiple deprivation). The four authorities are briefly characterised below:

- *Cardiff* – an urban area, typical in many ways of an administrative capital city with considerable polarisation in terms of education and income, and some ethnic diversity;
- *Bath and North East Somerset* – a mixed urban/rural area, including the city of Bath and the surrounding rural north-east area of Somerset. Polarised in education and income, and with the added advantage of having been well resourced in terms of public ICT access;
- *Blaenau Gwent* – Ex-mining and steel communities in the South Wales valleys with relatively impoverished levels of economic employment and education;
- *Forest of Dean* – a predominantly rural area, with high levels of poverty in some parts. The area has been used in previous studies as an English comparator for similar localities in South Wales. With a low population density and multiple deprivation, the area provides a key test area for virtual participation.

The next stage of the sampling process was to define the contours of the four study sites. Since most of the data used for initial selection was at the level of unitary authority and the electoral register was to be used to select the households, the sites were then defined in terms of electoral divisions (wards) within each authority. The criteria used to make selections of the three wards in each of the unitary authorities were educational achievement, poverty and population density – as reported in the Education, Skills and Training and Income domains of the Indices of Deprivation 2000 (IOD) and parliamentary electorate for 1998. Of the three wards selected from each of the unitary authorities, one was from either extreme (high/low) and one was more typical of both registers. Characteristics of the 12 wards are outlined in Table 3.2. The final column lists the sites available

Table 3.2 Descriptions of the 12 electoral wards

Name	Description	Population density	Income deprivation	Educational deprivation	Public ICT facilities
Ely (Cardiff)	City, housing estate	High	High	High	• Library with 11 open-access computers • Community centre with 5 open-access computers/also offers IT courses • Community Enterprise Centre with 5 open-access computers/also offers IT courses • Secondary school with computer learning centre • Methodist Church with 2 open-access computers
Canton (Cardiff)	City, mixed residential	High	Medium	Medium	• Library with 5 open-access computers • Cybercafé in city centre with 10 open-access computers • Adult Education Centre
Cyncoed (Cardiff)	City, affluent residential	High	Low	Low	• Library with 5 open-access computers
Cinderford (Forest of Dean)	Rural market Town	Medium	Medium	High	• Library with 2 open-access computers • Not-for-profit 'Telecottage' centre with 11 computers, ICT training, internet access and printing facilities • FE college Enterprise Centre, 2 computer suites, IT training
Tidenham (Forest of Dean)	Rural villages/ overspill communities from town	Low	Medium	Medium	• Learning centre in leisure centre in nearby town • Adult Education Centre in nearby town

Location	Community type				Facilities
Hartpury (Forest of Dean)	Rural villages and isolated small-holdings	Low	Low	Low	• UK Online centre in local FE college • Local FE college IT bus offering 'taster' ICT training sessions
Chew Valley (Bath and NE Somerset)	Rural villages	Low	Low	Low	• None in electoral ward
Lansdown (Bath and NE Somerset)	Affluent city	High	Low	Low	• City centre library with 26 open-access computers • Four authorised UK Online/learndirect centres • Three commercially-run cybercafés • One gaming cybercafé with 12 open-access computers
Radstock (Bath and NE Somerset)	Rural market town	Medium	Medium	Medium	• One gaming cybercafé with 9 computers • Library with 1 open-access computer • Local college 'shop front' learndirect centre
Nant-y-glo (Blaenau Gwent)	Mining community town	Medium	High	Medium	• None in electoral ward – although access to facilities in Ebbw Vale within 1–6 miles
Ebbw Vale North (Blaenau Gwent)	Mining community town	Medium	High	High	• Library with 4 open-access computers • Adult Education Institute with a 12 computer open learning centre, and IT training • FE college IT Learning Centre • One commercial cybercafé with 10 open-access computers
Beaufort (Blaenau Gwent)	Mining community village	Low	High	High	• None in electoral ward

for public access to ICT, garnered via websites, official publications, word-of-mouth and fieldnotes from site visits.

The household survey

The first tier of data collection involved a large-scale door-to-door survey of householders. Households in each electoral ward were identified from postal address files (the planned electoral registers were not all available). These lists were appropriate since the target population for the first wave was only those adults aged 21 years or more.

Systematic sampling, selecting every nth element after a random start, was an appropriate method to use but to avoid problems of periodicity and special methods of variance estimation, repeated systematic sampling was used. The sample was also stratified (in terms of age, sex and location) in an attempt to reduce the sampling variance and to ensure sufficient cases in certain categories, but the stratification was proportionate; i.e. the sampling fraction for each strata was uniform within wards. Reserve cases were pre-selected from adjacent postal addresses to cover non-response. The primary address was visited three times at different times of the day in successive weeks until contact was made with the householder. In a house with two or more householders, either was interviewed, according to quota and as convenient. If the house was clearly empty or all of the householders were out-of-strata, the first house on the reserve list became a new primary and the process started again. If the primary householder refused to take part or was not contacted after three calls, the first house on the reserve list was used instead, followed, if necessary, by the second reserve and so on. In the latter cases the response was recorded as a reserve for accounting purposes. The final sample comprised 1,001 adults, with a primary response rate of 75 per cent (Table 3.3). Within the sample, 41 per cent of respondents were male and 59 per cent female, 92 per cent were classified as 'white' and 8 per cent classified as 'non-white'. The age range of adults spanned 21 to 96 years with a mean age of 52 years (standard deviation 18 years).

In order to obtain a significant number of users of public ICT sites a purposive 'booster' sample of 100 users of public ICT sites in four of the wards was also surveyed. These data are used in our analysis of public centre usage in Chapter 8, resulting in a combined sample of 1,101 adults with 41 per cent male respondents, 91 per cent white British respondents and a mean age of 51 years (standard deviation 18 years). For all but specific analysis of the public ICT site data in Chapter 8, the initial sample of 1,001 adults was preferred due to its better claim to representation of the more general population.

As our project examines changes in learning histories and opportunities over a period of nearly 100 years, a longitudinal study allowing prolonged study of the lives of respondents would have many attractions but was impossible for practical reasons. Our use of retrospection also has some research advantages. It does not suffer from respondent attrition, or face the threat to internal validity coming

Table 3.3 Numbers of respondents in initial and boosted survey samples

	Initial sample size	Boosted sample size
Sex		
Male	405	44
Female	596	56
Age group (years)		
21–40	330	53
41–60	319	32
61 or more	352	12
Marital status		
Single/separated/widowed	355	38
Married/living with long-term partner	625	61
Health status		
No long-term illness/disability	229	12
Long-term illness/disability	761	88
Education		
Continued after 16 years	384	41
Completed education at or before		
16 years of age	617	59
Socio-economic status		
Service	83	5
Skilled non-manual	300	34
Skilled manual	87	13
Partly skilled	418	38
Other	113	10
Ethnic background		
'White British'	917	85
'Non-white British'	84	15
Household composition		
Single adult age 16–59	124	26
Small family	349	39
Large family	163	19
Large adult household	17	2
Adult aged 60 and over	192	11
2 adults, 1 or both aged 60 and over	156	3
Total	1,001	100

Notes
Data are number of respondents in sample (*n* = 1,001) and booster sample of public centre users (*n* = 100). Not all numbers will total 100% due to non-responses.

from the necessity to test and re-test the same individuals. Retrospective and longitudinal studies face problems of comparability over time (in the names of public qualifications, for example). However, such considerations are more problematic for a long-term study since the instrument to be used for all sweeps has to be designed before the changes that it needs to encompass. A retrospective study has the advantage of hindsight, and if it is used to collect relatively simple 'factual' data as ours is (rather than notoriously difficult to recall items, such as attitudes) then it is preferable.

A 36-page structured-interview instrument was developed, consisting of items covering detailed demographic details relating to the respondent and family, compulsory and post-compulsory educational histories, employment life histories and details of current and past ICT use at home, work and in community sites. In this way, the survey collected largely 'factual' information about respondents (the 'what' and the 'when'). The survey was administered by the researchers in co-operation with a university-based commercial research organisation in the summer and autumn of 2002. The interviews were held in people's houses, or infrequently by appointment elsewhere (e.g. place of work or relative's house). Being face-to-face, the method of delivery allowed interviewers to use show cards and read questions aloud, and allowed respondents to check answers with household records. This has several advantages, most notably by including those with limited literacy who are routinely invisible in postal surveys (Gorard 2003).

Methods of analysis

It was our contention that the household survey data were best analysed in a relatively straightforward manner. Thus, our analysis of much of the survey data is described in terms of frequencies, cross-tabulations and, where appropriate, means and standard deviations. However, in seeking to explain how patterns of educational participation and ICT use were potentially correlated with demographic characteristics such as age, sex or socio-economic status, a more sophisticated method of analysis was required. As Mossberger et al. describe, many studies seeking to describe such social patterning of ICT use and/or educational participation ignore the possibility of confounding or compounding variables:

> Most studies … report only simple frequencies, or per centages, which tell little about the strength of the association between variables or the relationship between variables. It is well-known that race, ethnicity, income and education are highly correlated, thus simple frequencies or per centages can overestimate the gaps in information technology based on any one of these factors. African Americans and Latinos, for example, tend to have lower incomes and educational attainment than do whites. The question, then, is whether race is an independent (non-spurious) predictor of access to information technology or whether, for example, education is really driving differences in access.
>
> (Mossberger et al. 2003: 16–17)

With this problem in mind, in Chapters 4 and 5 we present some of the results from a multivariate logistic regression analysis to 'predict' or 'explain' the various patterns of individual participation. In Chapter 4 the dependent variable in the regression represents four lifelong forms of participation in learning: 'non-participants'; 'transitional learners'; 'delayed learners'; and 'lifelong learners'. These patterns are also summarised in two binary variables – immediate post-compulsory and later-life participation. The independent variables, or potential determinants of participation, are entered into the regression model in batches in the order that they occur in the individual's life. This is instead of the more usual procedures of either entering all variables in one step, or stepwise in the order of the amount of variance they explain. The explanatory variables entered at birth were age, sex, place, and family background. The variables entered in the second stage were the nature of schooling, age of leaving full-time education, and first occupation. The variables entered in the third phase were modal occupational class, employment status, areas of residence, and own family. The variables entered in the fourth phase were the reported access to various technologies, including the internet. At each stage we also examined the impact of some of these variables in interaction – first occupation by sex, for example. In this way, the variables entered at each step can only be used to explain the variance left unexplained by previous steps, and are selected by using the likelihood ratio statistic. Thanks to this method of analysis, which models the order of events in individuals' lives, the relevant variables become valuable clues to the socio-economic determinants of patterns of participation in adult learning.

In presenting this analysis, and its equivalent analysis of ICT-based learners in Chapter 5, any redundant information in the tables is minimised for convenience. Only variables selected as possibly relevant by the modelling process, and retained using the likelihood ratio statistic are discussed. All models cited had a clear division between the two groups in terms of a predicted probability scattergram (although there are always a minority of cases mis-classified with marginal probabilities). The quality of the models in terms of goodness-of-fit to the data and log-likelihood were more than adequate for the analysis to proceed (for more on this, see Gorard *et al.* 1999).

In-depth semi-structured interviews

The second stage of the data collection involved in-depth, semi-structured interviews. The interviews were conducted by the researchers during the winter of 2002/3. Interviews were carried out with 100 respondents to the initial household survey. This sub-sample was purposively selected to include equivalent numbers of individuals with high/low levels of technology use and high/low educational background, with additional criteria of selection including age, socio-economic status, geography (urban/rural) and ethnicity. This resulted in an interview sample which (in terms of classifications derived from the survey data) could be categorised as 59 'frequent users', 15 'moderate users' and 26 current 'non-users' of computers (see Chapter 5). The demographic composition of the interviewees can be seen in Table 3.4.

All but three of the interviews took place in respondents' homes (two respondents indicating a preference for being interviewed in their workplace, and one finding it more convenient to visit the researchers' university to be

Table 3.4 Personal and demographic characteristics of interview sample

	Number of interviewees
Sex	
Male	47
Female	53
Age group (years)	
21–40	37
41–60	30
61 or more	33
Marital status	
Single/separated/widowed	24
Married/living with long-term partner	76
Health status	
No long-term illness/disability	75
Long-term illness/disability	25
Education	
Continued after 16 years	47
Completed education at or before 16 years of age	53
Socio-economic status	
Service	10
Skilled non-manual	39
Skilled manual	13
Partly skilled	30
Other	8
Total	100

Notes
Data are number of respondents in interview sub-sample (*n* = 100). Not all numbers will total 100% due to non-responses.

interviewed). As we specifically wished to elicit the voices of individual adults these interviews were usually conducted without the presence of other family members. Interviews lasted between forty minutes and one-and-a-half hours. The common interview schedule focused on individual's domestic, educational and employment 'careers' as well as their technological histories and their present technological activities. In this sense, the interviews approached a life history or life story method in that they were focused on eliciting an individual's experiences through a chronological autobiography of home/family, education, work and technology use (see Dhunpath 2000). Obviously people's use of technology and their engagement in education are complex affairs and, in fact, are inevitably less straightforward than many of the elicited narratives from our interviewees which tended to present a coherent life-story (McAdams 1998). Nevertheless, these semi-structured interview data do allow for a more detailed investigation of adults' perceptions of the factors influencing their use (and non-use) of ICT for learning, as well as their participation in lifelong learning. They complement the 'what' and the 'when' of the survey details for the same individuals with more of the 'why' and 'how'.

Methods of analysis

As explained at the start of the chapter, one of the major advantages of our staged approach to the study is that the large-scale patterns, and some of the socio-economic determinants, of adult participation in learning are known by the time of the interview stage. This means that many of the preliminary codes used in analysing the interviews already exist. These included the determinants themselves (such as the way in which the experiences of place or parental family are described in relation to learning or ICT-use), and stories of key decisions or transitions. In addition, of course, the interviews themselves generated further codes during analysis, and these suggested further ways of examining the household survey data. Thus, the overall process was iterative rather than strictly staged.

Case studies

Finally to gain detailed insights over time into how individuals are making use of ICTs and how ICTs fit into their day-to-day lives, in-depth case studies of individuals and their associated households were selected from the 100 second-stage interviewees. Each case study involved three home visits including a series of semi-structured interviews with the respondent, one group interview with other family members, observation of ICT use by the case study individual, and documentary analysis. Thus, a variety of methods were used in these 'ethnographic case studies':

- in-depth interviews repeated over a twelve month period;
- observations during interviews – including the respondent taking the researcher on a 'tour' of their computer and demonstrating their favourite or most used programmes, websites and applications;
- mental mappings of the house (to visualise the physical space in which ICTs were placed);
- network diagrams (to make visible the social networks these households were embedded in);
- photographs of the respondents' houses and ICT equipment.

A similar range of methods has been used in other studies with children to explore their use of technologies such as the television, computer and internet (e.g. Facer *et al.* 2003; Tally 2004). As Frissen explains, such an approach affords a rich understanding of often unconscious activities and behaviours:

> the purpose of this combination of methods is to make people talk about their everyday life and their everyday use of ICTs in their own words, which enables us to reconstruct more or less routinised and taken-for-granted patterns of behaviour. Triangulation of methods is important, because other research has shown that it is not easy for respondents to talk about their everyday use of media and ICTs and their underlying communication needs, and especially

their future needs. Contextualising uses and needs as much as possible is a way to solve this problem.

(Frissen 1999: 66)

Twenty-five respondents were purposively selected for these year-long case studies from the sub-sample of 100 interviewees. All of the sample were moderate or high users of ICT. Other criteria for selection included educational background, age, class, geography (urban/rural), household composition and ethnicity with a view to forcing variation to encourage a range of accounts. Table 3.5 summarises the key features of each family that we worked with over the 12 month course of the case studies.

Conclusion

In this chapter we have described the rationales behind our research questions and subsequent research design. This represents a decent but, as with all social research, imperfect attempt at mapping ICT and adult education in its many forms. Having completed the project our triangulating methods allow us to reflect on the accuracy of all the phases of data collection. It was apparent that it was not until the second or third case study visit (i.e. the fourth or fifth time which they had been part of a data collection process) that some respondents began to feel comfortable enough with the research project and ourselves as researchers to 'open up' about their (non)use of ICT. When this happened, there were sometimes telling incongruities between the initial survey data and our later qualitative data – from the initial under-reporting of people's ages to over-reporting of technology use and engagement with learning. It was sometimes not until the final case study visits that the differences between people's 'official lives' as first reported to the survey administrators and their actual lives became apparent. Firmly stated intentions to take courses, buy computers or move house had never quite materialised when we next visited, however imminent or strongly stated the intentions were. Of course, this has implications for all other work of a 'snapshot' nature.

The fact that people have learning and technology 'careers' which ebb and flow alongside the rest of their lives was sometimes painfully apparent in our data collection process. People's stated good intentions were altered by often unforeseen life events over the course of the project, such as immigration, divorce, death in the family, career progression or events connected with children. At all times we sought to merely document what we were told in response to enquiries. Although our data, as with all social science data in general, may not be an accurate reflection of the 'truth', it does faithfully document people's perceptions, which is as good as can be hoped for.

It is also worth reflecting briefly on the extent to which our research activities involved an element of reactivity amongst some respondents – allowing them to reflect upon their use of ICT and sometimes act as a stimulus for further

Table 3.5 Characteristics of the case study respondents

Name	Occupation	Sex	Age	Location	Household	Education background	ICT background
Burton	Skilled manual	M	77	Cyncoed	2 adults aged 60+	1–4 O-levels/GCSEs	>5 years experience
Dobson	Skilled manual	M	39	Canton	Small family	1–4 O-levels/GCSEs	>5 years experience
Fitzgerald	Skilled non-manual	M	34	Canton	Small family	1–4 O-levels/GCSEs	>5 years experience
Frampton	Partly skilled	M	41	Cinderford	Large family	1–4 O-levels/GCSEs	>5 years experience
Hargreaves	Skilled manual	F	51	Hartpury	Adult aged 16–59	No qualifications	<5 years experience
Harrold	Unskilled/other	F	30	Tidenham	Small family	A-levels	>5 years experience
Hunt	Unskilled/other	F	33	Canton	Small family	5 or more O-levels/GCSEs	>5 years experience
Hutchinson	Skilled non-manual	F	39	Beaufort	Small family	No qualifications	>5 years experience
Julian	Skilled non-manual	F	57	Ebbw Vale	Large family	Degree	<5 years experience
Lawrence	Skilled non-manual	F	45	Canton	Large family	1–4 O-levels/GCSEs	>5 years experience
Lennie	Service	M	61	Beaufort	2 adults aged 60+	5 or more O-levels/GCSEs	>5 years experience
Myers	Unskilled/other	F	22	Cyncoed	2 adults aged 16–59	No qualifications	>5 years experience
Nelson	Unskilled/other	F	29	Ely	Small family	No qualifications	<5 years experience
O'Connor	Unskilled/other	M	31	Cinderford	Small family	1–4 O-levels/GCSEs	>5 years experience

continued…

Table 3.5 continued

Name	Occupation	Sex	Age	Location	Household	Education background	ICT background
Palmer	Skilled manual	M	55	Canton	2 adults aged 16–59	No qualifications	<5 years experience
Peters	Partly skilled	M	62	Lansdown	2 adults aged 60+	Degree	<5 years experience
Rankin	Partly skilled	F	34	Radstock	Large family	No qualifications	<5 years experience
Rhodes	Skilled non-manual	M	33	Lansdown	Large family	No qualifications	>5 years experience
Salako	Skilled non-manual	M	57	Hartpury	Small family	No qualifications	>5 years experience
Smith	Partly skilled	M	31	Beaufort	Adult aged 16–59	No qualifications	>5 years experience
Sodje	Service	F	54	Hartpury	2 adults aged 16–59	Degree	>5 years experience
Somner	Skilled non-manual	F	36	Cyncoed	Small family	A-levels	>5 years experience
Tabb	Service	M	50	Cyncoed	Large family	5 or more O-levels/ GCSEs	>5 years experience
Talbot	Unskilled/other	F	28	Radstock	Small family	A-levels	>5 years experience
Turner	Service	F	50	Chew Valley	Small family	Degree	>5 years experience

Note
Names of respondents are pseudonyms.

engagement. Just as Elaine Lally's (2002) interviews with families were later acknowledged by her subjects as being instrumental in precipitating their purchasing of computers, our visits were also acknowledged by some respondents as spurring them into action. This issue notwithstanding, taken with the usual methodological caveats we feel confident that we have been able to collect a quantity and quality of data to do justice to our research aims and questions. With this in mind we now go on to present our analysis of the survey and interview data – first in terms of patterns of participation in education and learning.

Chapter 4

What makes a lifelong learner?

Introduction

The next two chapters map the patterns and determinants of adults' participation in education and use of ICT. As such, they provide the foundations for our empirical understanding of the le@rning society. Using the data from our household survey, they describe what learning is taking place, to what extent this learning involves ICT, and what activities ICTs are being used for if not learning. The level and nature of adults' engagement with ICTs is more fully explored in Chapter 5. The present chapter is concerned with describing and discussing the patterns of formal educational participation evident in our data.

As we saw in Chapter 1, the notion of lifelong learning has recently awakened new interest amongst politicians and the education community, for a variety of reasons. Among these reasons are an apparent decline in 'jobs for life', leading to a demand for retraining, transferable skills, multi-skilling and 'careership'. There is a perceived need to invest in human capital to compete effectively for local inward investment in a globalised economy, and also the advocacy of greater adult participation in education and training to combat social exclusion and increase equity. This has led to considerable policy-talk about the need to encourage wider, longer and more regular participation in relevant learning among adults. As we have discussed, one of the central tenets of the drive towards widening participation in adult learning lies in the facilitation of easy access to learning resources and opportunities away from the traditional confines of educational institutions. The use of ICTs is widely regarded as a key means by which this goal will be accomplished – something we have labelled the 'le@rning society' thesis. Therefore, alongside the potential economic benefits of up-skilling the workforce, technology-based learning is also being enthusiastically promoted as a new way of combating social exclusion.

In order to test these assumptions empirically the present chapter focuses on the patterns of formal learning apparent in our household survey of 1,001 adults. Where appropriate, data from our in-depth interviews with 100 of the initial respondents are also used for illustrative purposes. Using simple frequency counts, cross-tabulations and the logistic regression technique outlined in Chapter 3, we address the following questions:

- What are the patterns of participation in lifelong learning?
- What are the determinants of participation in adult learning?
- To what extent does use of ICT interrupt or reinforce existing patterns of participation in lifelong learning?

What are the patterns of participation in lifelong learning?

Of the 1,001 adults in our household survey only 38 per cent continued with any form of formal learning directly after reaching compulsory school-leaving age – what we shall term from now on 'immediate learning'. Only 46 per cent reported any formal learning apart from that directly after reaching compulsory school-leaving age – what we shall term 'later learning'. The forms that these patterns of immediate and later learning took for our survey respondents can be illustrated by categorising them into one of four learning 'trajectories' representing different lifelong forms of participation:

- 'non-participants' are those who reported no episodes of education or training at all since leaving school at the earliest opportunity;
- 'transitional learners' reported at least one episode of immediate post-compulsory education or training but nothing subsequently in later life;
- 'delayed learners' reported no episodes of immediate post-compulsory education or training but at least one subsequent episode as an adult in later life;
- 'lifelong learners' reported at least one immediate episode of post-compulsory education or training and at least one other episode in later life.

As can be seen in Table 4.1, the most common pattern within our survey was a report of no formal education or training at all since reaching compulsory school-leaving age – the 'non-participants'. Our finding that over a third of respondents reported non-participation in any adult learning whatsoever is similar in scale to the figure reported by La Valle and Blake (2001) – 41 per cent of whose respondents reported no taught learning in the prior three years. Significantly, our survey data also mirror many others in finding that participation is patterned by sex, age, ethnicity, disability, caring responsibilities, educational background, employment, and local deprivation. The exact patterning by each demographic characteristic is now discussed in turn.

The influence of age and time

Table 4.1 suggests considerable disparity in patterns of participation by mean age of respondent. This is particularly powerful, since it highlights clearly how older groups are less likely to have been involved in *any* learning, despite the longer time they have had to do so. Table 4.2 shows the changes over time in each age cohort. Non-participation has declined, and has partly been replaced by

Table 4.1 Frequency of four patterns of participation

Trajectory	Frequency of trajectory	Percentage of that trajectory	Mean age of that trajectory, in years
Non-participants	371	37	58
Transitional learner	175	18	45
Delayed learners	246	25	52
Lifelong learners	209	21	44

Note
Some columns may appear not to sum to 100 per cent due to rounding.

Table 4.2 Patterns of participation by age range

Trajectory	Aged 21–40	Aged 41–60	Aged 61+
Non-participant	26	28	55
Transitional learner	23	18	11
Delayed learner	22	29	24
Lifelong learner	29	25	9

Note
The cells contain the percentage of each trajectory within each age range. Some columns may appear not to sum to 100 per cent due to rounding.

transitional learning, as the expectations of what was 'normal' have also changed over time. A story that appeared in various forms among our interviewees in the older cohort involved leaving school at the earliest opportunity, because that was considered to be the 'norm' and also because there were jobs or family responsibilities to be taken on. In the second example below, this led to frustration and therefore, perhaps, to a later return to formal learning.

> I was only 14 [when I left school] ... it was just the primary school in the village, you didn't go on to higher education [then] unless you won the 11+ scholarship ... I [tried and] failed ... I was brought up on the farm and my father said, well there were three girls and one boy, and he said 'well, one of you will have to stay home and help your mother, so it may as well be you'.
>
> (female, older, non-participant)

> My dad was very Victorian or Edwardian or whatever, and his theory was that girls left school, got a job, and then got married and had children ... He was very reluctant to push us into higher education, which is what I should have done.
>
> (female, mid-age, delayed)

Jobs were easy to get, and there is an implication in these stories that, in such times, staying on in education has no inherent merit. The next interviewee also mentions a very common theme among older and mid-age respondents – that of learning by doing and common sense. As this example demonstrates, quite considerable changes in skilled tasks have been undertaken in the past without any formal training:

> I didn't leave with nothing [qualifications]. I left school on the Friday, not on the Thursday, because my birthday was on the Friday, and I started work on the Wednesday because it was Easter ... everybody went in to get an apprenticeship then, you know ... nobody stayed on after fifteen then, unless you was in grammar school ... there was plenty of vacancies in boot and shoe [factory]. And plenty of vacancies in engineering.
> [...]
> I went down for a driving job and the bloke said to me 'can you drive?', so I said 'yeah' ... 'Have you got a licence?' I said 'yeah'. 'Can you start Monday?' I said 'yeah'. And I got there on the Monday and the biggest thing I drove then was a [car], and I got there on Monday and there was this bloody great pantechnicon. He said 'There you are, you're off to London'.
>
> (male, older, delayed)

Our survey data showed that 'transitional' learners are more common among the younger and mid-age respondents. This is partly due to the increase in post-16 educational opportunities and the expectations, both parental and societal, that went with them. However, this does not in itself lead to an increase in lifelong learning, perhaps the reverse, as explained by the following respondent.

> I hated every minute of it ... I thought it was the most boring waste of time ... I stayed in the sixth form because I was expected to ... nobody I knew left school at 16 ... well I met my husband when I was 17. [On applying for jobs] They all wrote back and offered me jobs. Because at that time if you had two arms and two legs and could write your name you were in, you know.
>
> (female, mid-age, transitional)

Although the potential learning experiences of 21–40-year-olds are most clearly truncated by their relative youthfulness, the difference between transitional learners and the others is not simply one of age. This is evidenced by the fact that lifelong learners are, on average, the youngest. Therefore, at least part of the assumption inherent in the learning society model that individuals will choose to continue to build upon their prior learning throughout the life-course is not borne out. In fact, the massification of further and higher education in the UK for the two youngest age cohorts has to some extent simply replaced later learning (and delayed participation) by front-loaded immediate education (and transitional participation). Over the past 60 years, compulsory schooling has been extended from age 14 to age 16, and staying-on rates for further and higher education after that age have increased considerably. Therefore, the frequency of non-participants has dropped over the same period. More people are now likely to undertake some formal education or training post-16, but most of these undertake it only immediately after reaching school-leaving age. The greatest growth has been in transitional learners. While lifelong learners have increased over time, delayed learners have correspondingly decreased a little. These changes largely reflect differences in patterns of participation for women (see below). The replacement over time of later work-based training by increased further education when young, that we observed, has also been suggested by Denholm and Macleod (2003).

The influence of place

Table 4.3 is the first of several tables which illustrate the importance of place in patterning access to learning opportunities. It shows that participation was found to be weaker in Blaenau Gwent, an area with considerable socio-economic disadvantage, and higher in Bath and North East Somerset, an area which includes the prosperous tourist city of Bath. As outlined in Chapter 3, these areas were selected as study sites due to their differing social and economic circumstances. Therefore, we can assume that this variable is acting as a proxy for relative social advantage among the local population, as well as pointing to the varied nature and scale of local opportunities for adult learning. As this interviewee from one of our South Wales valleys communities indicates:

> There is nothing in the valleys now, nothing at all … It's basic employment in the valleys now. You've got to be in the cities to get the good jobs. I mean it is run of the mill on the council and things like that but I wanted my children to have a bit more than what I had.
>
> (female, mid-age, non-participant)

Table 4.4 contains one of a number of indicators that show how participation is strongly related to geographic mobility. In general, participation increases with the distance between current area of residence and area of birth. The most mobile (i.e. those who have moved house the most), and those born furthest from the neighbourhood where they were surveyed, are the least likely to be non-participants. In these explanatory extracts, the first example is of a non-participant living in the same rural area all of her life, and the second is of a lifelong learner showing that the link between education and mobility is a complex one.

> We lived in a sort of well-out-of-the-way place in the woods. So we had to walk a long way to school anyway. So if the weather was really bad and there was snow up to here, that was another good excuse to stay home … I think moving out of that environment and going to school, there's so many people, it's so annoying, see.
> […]

Table 4.3 Patterns of participation by area of residence

Trajectory	Living in Forest of Dean	Living in Bath/ NE Somerset	Living in Blaenau Gwent	Living in Cardiff
Non-participant	35	25	53	36
Transitional learner	12	26	9	23
Delayed learner	31	22	26	20
Lifelong learner	23	27	12	22

Note
The cells contain the percentage of each trajectory within each area. Some columns may appear not to sum to 100 per cent due to rounding.

Table 4.4 Patterns of participation by area of birth

Trajectory	Born in current ward	Born in current district	Born in current area	Born in UK	Born abroad
Non-participant	50	47	24	24	14
Transitional learner	17	12	19	21	24
Delayed learner	23	29	30	20	24
Lifelong learner	10	13	26	35	38

Notes
Ward is defined here as the same electoral ward as where the interview took place; district is the same local authority; and area refers either to South Wales or the west of England.
The cells contain the percentage of each trajectory born at each distance. Some columns may appear not to sum to 100 per cent due to rounding.

I [left school at 14 and] went to work in Woolworths … They only just, you know, showed you what to do and you should have got on with it. People didn't have any special training as such.

(female, mid-age, non-participant)

The problem was that my family ran – we did pubs – so we were moving constantly … what we used to do was buy a pub … redecorate it, revamp it, refurbish the whole thing … then sell it on and go on to the next one … the big issue was moving constantly, because you'd move from school to school… so my education went up and down, so I didn't do as well in my O-levels as I could have done, and I know that, and I think that's what probably spurred me on.

(female, younger, lifelong)

The influence of sex, ethnicity and religion

Table 4.5 does not show powerful differences overall between males and females (although the difference in terms of delayed learning shows up later as a key element). It does show, however, that those from minority ethnic backgrounds are more likely to participate, both as transitional and lifelong learners.

Table 4.5 Patterns of participation by personal characteristics

Trajectory	Male	Female	White British ethnic origin	White other ethnic origin	Other ethnic origin
Non-participant	34	39	37	45	23
Transitional learner	18	17	16	28	35
Delayed learner	27	23	26	7	15
Lifelong learner	20	21	21	21	27

Note
The cells contain the percentage of each trajectory within each category. Some columns may appear not to sum to 100 per cent due to rounding.

Respondents from Catholic and minority religions (including Moslems and Hindus) were also more likely to report some form of participation than those from Anglican families and those with no religion. These variables, and the others that follow are therefore used as potential determinants of participation in our later regression model.

The influence of parental family

There are also clear relationships between participation and the characteristics of respondents' parents. In general, the patterns are the same for father and mother, and only a selection of results are shown here. These relationships are not new, and have been remarked before by a variety of commentators (e.g. San-Segundo and Valiente 2003; Gorard et al. 1999). Post-compulsory ('immediate') participation for the respondent is reflected in the elevated age of leaving education for the mothers and fathers of transitional and lifelong learners (Table 4.6). It would appear that there is considerable reproduction over generations in terms of learning experiences and qualifications. Only 2 per cent of those whose mother has a degree are themselves non-participants, whereas 42 per cent are lifelong learners, for example. We do not consider here the qualifications of the respondent, since in the model described below their patterning was seen to be a proxy for prior determinants (i.e. while participation in education and subsequent qualification *are* related, they are both equally predictable from prior background characteristics).

The following example from our interviews illustrates the 'reproduction' of education patterns within families, and the power of initial schooling, with family

Table 4.6 Patterns of participation by parents' education

Trajectory	Mean age father left school, in years	Mean age mother left school, in years
Non-participant	14	14
Transitional learner	16	16
Delayed learner	15	15
Lifelong learner	16	16

influence, to help create a lifelong attitude of confidence in face of objective learning opportunities:

> It was a fee-paying school, a private school. There wasn't anything different or special. It was just a normal school [in Australia]. My father attended this school, so it was pre-ordained that I would attend this school as well ... I enjoyed it very much. It was a good school ... It was kind of, sort of, ingrained that we would be conscientious and do the best we could. They were not the type of parents to push things. At the same time, I guess, they led by example. They were fairly well-learned themselves ... my father's an orthopaedic surgeon and did a lot of study to get that far. My mother... she was a young children's teacher for a while ... I think it set the basis for confidence ... I'm not scared by anything.
>
> (male, younger, lifelong)

This reproduction was not always welcomed by its recipients, as this younger woman with two degrees funded by her father reflects:

> I didn't get a job for a while and I couldn't think of anything that I wanted to do and then my father led me into doing a masters. Because my father had led me into it and was paying for it I ended up doing a subject that interested him rather than one that interested me, that also sounds stupid that I have two degrees in subjects that I'm not particularly interested in, my masters degree is in transport.
>
> (female, younger, lifelong)

The influence of social class in these patterns of reproduction is also borne out from our survey data. Table 4.7 shows that, for example, respondents from families with non-manual occupations or service class (professional/managerial) fathers are substantially more likely to continue to further education or training than those from families where fathers do not work or are part/unskilled. Family background is therefore one of the most important predictors of participation (see below).

Table 4.7 Patterns of participation by father's social class

Trajectory	Father service class	Father non-manual job	Father skilled job	Father part-skilled job	Father unpaid
Non-participant	5	13	24	49	55
Transitional learner	40	27	20	12	18
Delayed learner	18	22	29	26	18
Lifelong learner	37	39	28	14	8

Note
The cells contain the percentage of each trajectory within each class. Some columns may appear not to sum to 100 per cent due to rounding.

> I come from a very rural area in West Cornwall. My father was a tin-miner ... we were a bit poverty-stricken.
>
> (male, older, delayed)

> My father was very keen for me to succeed at school and put a lot of pressure on me ... I think my father was disappointed I left. I did not really think about a career, I just wanted a job ... I began in the lab testing steel quality. I was not trained – simply told to go and look at someone else and then get on and do it myself.
>
> (male, older, delayed)

Table 4.8 relates also to Table 4.4 in showing the link between mobility and learning, and therefore the importance of place. Participation for respondents increases with the distance between their current area of residence (the study site) and the area their mother, or father, was born in. We develop this point later.

The influence of compulsory education

Respondents who reported not attending compulsory school regularly were less likely to also report adult participation in learning of any sort (Table 4.9). Indeed 60 per cent of self confessed 'non-attenders' reported no adult education or training at all (and this despite their average age of 52). It is notable, however, that 22 per cent of them did report a return to some formal learning at a later date.

As might be expected, many more respondents with a background of private or selective education had become lifelong learners (using the current definition) than those attending other state-funded schools. These patterns are clearly determined historically, geographically and socio-economically as much as educationally. For example, attending a private school could be a proxy variable for many of those above (such as having a father in a service class occupation). Nevertheless, despite the importance of early family background, experience of initial schooling was found to also be part of the pattern set for later life participation. As some interviewees explained:

> Hated it. I didn't like school at all ... I used to play up a lot ... It wasn't relevant to the job I got in the end. I was lucky really.
>
> (male, younger, delayed)

> [On passing the 11+ exam] So, yes, my schooling played an important part in my future – I didn't know it then but I know now with the benefit of hindsight ... I didn't leave with any qualifications ... I walked into a job. See, that was a different thing then ... Half the class I was in had jobs before they left school ... Engineering. Most in that school went into engineering.

Table 4.8 Patterns of participation by mother's place of birth

Trajectory	Mother born in ward	Mother born in district	Mother born in area	Mother born UK	Mother born abroad
Non-participant	56	49	26	23	25
Transitional learner	12	13	22	21	23
Delayed learner	23	27	27	23	16
Lifelong learner	9	11	25	32	36

Notes
Ward is defined here as the same electoral ward as where the interview took place; district is the same local authority; area refers to South Wales or South West England.
The cells contain the percentage of each trajectory for each distance. Some columns may appear not to sum to 100 per cent due to rounding.

Table 4.9 Patterns of participation by attendance at school

Trajectory	Regular school attender	Not regular school attender
Non-participant	35	60
Transitional learner	18	10
Delayed learner	25	22
Lifelong learner	22	8

Note
The cells contain the percentage of each trajectory within each category. Some columns may appear not to sum to 100 per cent due to rounding.

[On not being allowed to sit exams at 16] I was offended more than anything. I mean, I wasn't a brilliant student ... but then as an apprentice my employer sent me to day release and then I got qualifications.

(male, older, delayed)

My sister went to grammar school and I went to a technical school ... I would have liked to [go to the grammar school] but I wasn't as clever as she was ... my father was a teacher and my mother didn't work at all. So it wasn't really a big deal that I didn't go to the school that she went to. No. Not really. It's just that I didn't like school. But I did more night school later on. [...]
I just watched her. She would tell and show me. That's all really ... So I was then doing calf-rearing, milking, sheep... Just by people showing you, the farmer showing you. I loved it. I didn't find it hard.

(female, older, delayed)

The influence of current family set-up

The respondents' own family set-up was often found to be a key indicator of later learning patterns, especially in interaction with respondents' sex (and occupational class). For men, living with a partner did not seem to be the barrier to participation that it can be for women; in fact it was linked in our survey data to enhanced participation. A similar pattern applies also to having children to look after. The relatively high proportion of delayed learners among those who have children suggests the respondents' learning ambitions were initially frustrated by the need to care for the children (see Table 4.10). Indeed, in the semi-structured interviews, there were several stories from female interviewees about the clash between family commitments and the desire to take part in formal learning episodes:

> Yes, she was pushing me to go back, but with the children – it's not easy at all you know. I've got mum and dad as well with me. My dad's got Alzheimer's disease … maybe later, not now.
>
> [...]
>
> After I had my daughter, I wanted to know all about the way I could play with her, the way I can teach her, the way I can enjoy things with her. And I absolutely loved it.
>
> (female, younger, transitional)

> I couldn't wait to get out. I got out of there early, and I left when I was 14 … I used to sign in and naff off … I don't think I was put on this earth to be told what to do.
>
> [...]
>
> I thought I'd put it on a back burner for a later date, then I had Cory and Charles and it's just the way it goes … I would go back now if I could.
>
> [...]
>
> Anything at all. I love to read. I read biographies and thrillers, anything … yes always, I could read when I was four, so I always like to.
>
> (female, younger, non-participant)

Although these women had been successful at school, and had clear plans about how their career and life was to have unfolded, the job was often put on hold perhaps because of children and so they 'essentially replicated the employment patterns of women of an earlier generation' (Aveling 2002: 265). What both of these extracts also exemplify is that lack of participation in formal episodes

Table 4.10 Patterns of participation by current family set-up

Trajectory	Single	Have partner	Have children	Have no children	Mean number of children
Non-participant	49	30	36	29	1.6
Transitional learner	17	18	15	24	1.2
Delayed learner	21	27	29	15	1.9
Lifelong learner	13	26	20	32	1.5

Note
The cells contain the percentage of each trajectory within each category. Some columns may appear not to sum to 100 per cent due to rounding.

was often not only the result of lack of motivation, or the lack of opportunities. Sometimes personal circumstances are a hindrance, but they are often also the start of another, less formal, route into learning by reading or child-care. Our final example provides a reverse gendered perspective, with this man now conforming to his wife's notion of a wage-earning 'family man' rather than pursuing the less lucrative pathway of postgraduate study:

> [My wife] said to me, she said 'go on, get your degree' she said, 'if it works out for you fair enough', she said 'but one thing that you've got to remember is that you'll be a family man then and you can't go straight on [to postgraduate study] – because you'll have to pay back a loan. Let's see if we can get you a job before you go on to the next stage'. And so at the minute I'm stuck ... I want to progress my degree and I want to use my degree.
>
> (male, younger, delayed)

The relationship between formal and informal adult learning

Our study also asked a series of questions about informal, leisure and self-directed learning of a substantive nature (i.e. an interest that was sustained for at least one year). This is a tremendously difficult area to research accurately via a survey, but Table 4.11 summarises a simple notable difference between those who did and did not report such an interest. Perhaps two points are worthy of special note. First, this kind of informal learning is patterned in a similar way to formal learning. Second, however, there are substantial numbers of apparent 'non-participants' in formal learning who reported sustained interest in informal study.

What are the determinants of participation in adult learning?

Putting all of the above together in a multivariate analysis, we are able to 'predict' using logistic regression which of the four 'trajectories' is reported by each respondent with considerable accuracy using only what we know about their non-educational or initial education background (e.g. year of birth, sex, father's

Table 4.11 Patterns of participation by report of informal study

Trajectory	Reported informal study	Not reported informal study
Non-participant	25	41
Transitional learner	17	18
Delayed learner	28	23
Lifelong learner	31	18

Note
The cells contain the percentage of each trajectory within each category. Some columns may appear not to sum to 100 per cent due to rounding.

occupation, and type of school attended). For example, we can predict whether any individual will have reported an immediate period of extended initial education or training with 84 per cent accuracy (i.e. our prediction would place the individual on the correct trajectory 84 per cent of the time). By the time we predict their later-life episodes of participation, the overall accuracy is 77 per cent (Table 4.12). In both cases, therefore, we improve the accuracy of our predictions by around 50 per cent compared to chance. More significantly, by creating the model in a hierarchical way, we calculate it in terms of explanatory variables entered in batches representing periods in the individual's life from birth to the present. In this way, each batch of variables can only add to the best prediction based on the previous batch(es), and only 'explain' the variance so far left unexplained. This gives us a clue as to which variables are the determinants of learning episodes and which, like qualifications (see above), are simply proxy summaries of the others.

As can be seen in Table 4.12, the vast majority of variation in patterns of participation that can be explained is explained by variables that we could have known when each person was born. Other than that the key issue in explaining continuous post-compulsory learning is the experience of initial schooling, whereas the key issue in explaining later life learning is experience of work and family life as an adult.

Tables 4.13 to 4.16 show the coefficients for the logistic regression model underlying the predictions in Table 4.12. As noted in the previous section, even when the interaction with other variables is accounted for, younger respondents are more likely to continue with initial education and training of some sort, as were those from a minority ethnic background (Table 4.13). People aged 21–60 years were nearly three times as likely to take part in later education or training as those aged 61 years or over, and so on. An interesting change occurs in patterns by ethnic group. Respondents of minority ethnic origin were considerably more likely than those in the 'white British' category to continue education and training past school-leaving age, but considerably less likely to participate later in work-based training or adult education classes.

Males were almost twice as likely as females to take part in later learning. However, in our previous work (Gorard and Rees 2002), when the respondents were divided into age cohorts there was a significant complication. In this earlier study, we found that among those aged 38–65 years, men were three times as likely as women to have completed further study immediately after school, but no more likely to have undertaken any study thereafter. Among those aged 21–37 years, men were no more likely than women to have completed further study after school, but over four times as likely to have undertaken study thereafter. These figures suggested that while initial post-compulsory participation has become gender neutral, later participation now over-represents men in a way that it did not before.

Returning to our own data, the incidence of work-based training was found to have declined over the past 50 years (although there is no evidence that

Table 4.12 Percentage of patterns of participation correctly predicted at each life stage

	Variables at birth	Variables at end of schooling	Variables as adult
Immediate learning	79	84	84
Later life learning	71	71	77

Note

The use of information from adult lives cannot be said to 'predict' episodes of continuous initial education anyway.

Table 4.13 Personal characteristics as determinants of participation

	Immediate	Later life
Year born	*1.03*	
Ethnic group		
White British	0.57	2.24
White other	1.24	0.82
Other ethnicity	–	–
Where born?		
Same ward	0.56	
Same site	0.62	
Same region	0.81	
UK	1.66	
Overseas	–	
Sex		
Male		1.95
Female		–
Age group		
21–40		2.90
41–60		2.81
61+		–

Note

For clarity, real-number variables have been written in italics meaning that the reported coefficients are multipliers for that variable. For example, someone born in 1975 is 1.03 times as likely, *ceteris paribus*, to continue in education or training after school as someone born in 1974. All other coefficients are for categorical variables, and represent a change in odds compared to the final category. For example, a male is, *ceteris paribus*, 1.95 times as likely as a woman to undertake an episode of education or training in later life.

employment episodes themselves are any shorter now than they were in the 1950s). Indeed, not only is training declining somewhat in frequency but those episodes now reported are notably shorter on average (reflecting the growth of ICT and health and safety courses perhaps). This means that employers have been funding, directly and indirectly, a declining share of adult learning over time. Perhaps one reason that later learning is less common for women is that almost no training takes place in industries with high proportions of 'flexible' (i.e. short-term and part-time) labour. Women are currently over-represented in these industries, and it is these, not learning organisations requiring highly-skilled personnel, that have actually been the growth area over the last decade. Coupled with the near demise of uncertificated adult education classes in the UK it is clear that later learning is in nowhere near as healthy a state as initial education. Also apparently in decline is the incidence of substantial informal learning, undertaken by those who were otherwise not participating. Current training and certification arrangements do not recognise the relevance of these informal episodes.

As noted above, the influence of parental background is also significant (Table 4.14). The age that parents finished initial education was found to be a key determinant in our dataset. Also a respondent from a family with a service class

Table 4.14 Parental characteristics as determinants of participation

	Immediate	Later life
Age father left education	1.14	
Age mother left education	1.28	
Father's occupation		
Service class	3.82	2.87
Non-manual	3.06	3.40
Skilled manual	2.01	3.22
Part-skilled	1.18	2.12
Unpaid	–	–
Mother's occupation		
Service class		0.82
Non-manual		1.51
Skilled manual		0.14
Part-skilled		1.21
Unpaid		–
Father's residence		
Same ward		0.62
Same site		1.07
Same area		2.06
UK		1.43
Overseas		–
Mother's birthplace		
Same ward	0.50	
Same site	0.52	
Same area	1.61	
UK	1.25	
Overseas	–	

Note
For clarity, real-number variables have been written in italics meaning that the reported coefficients are multipliers for that variable. For example, someone whose mother left initial education at age 16 is 1.28 times as likely to continue in education or training after school as someone whose mother left school aged 15. All other coefficients are for categorical variables, and represent a change in odds compared to the final category. For example, someone whose father had a service class occupation is 3.82 times as likely to continue in education or training after school as someone whose father was unpaid (unemployed, or economically inactive).

father is nearly four times as likely to continue with education, and nearly three times as likely to undertake formal learning in later life, as one from a family with a non-working father. The picture for mothers is very different. Once other factors are taken into account, the occupation of mothers is unrelated to immediate post-compulsory learning. But a respondent with a non-working mother is nearly seven times as likely to report later life learning as one with a mother from a skilled manual occupation. This, presumably, reflects the middle-class family structure with traditional gendered roles, as much as any learning advantage from having a non-working mother *per se*.

Table 4.15 shows the model coefficients for relevant variables relating to schooling, and then initial occupational class. Those respondents who did not attend school regularly were only one quarter as likely to continue to education or training at school-leaving age as the others. Those attending private or grammar schools were considerably more likely to continue learning at school-leaving age than those attending elementary, and other, schools. This pattern is partly historic reflecting the decline of elementary schools but, given that year of birth has already been accounted for, it also reflects the 'social-sorting' impact of the school system in England and Wales. While occupational class immediately on leaving education is strongly linked to both continuous and later learning, the patterns are different. Whereas service class respondents were more likely to have continued with full-time education (unsurprisingly), later learning was even more common for the non-manual and skilled manual occupational groups (in comparison to those without paid employment) reflecting the greater prevalence of work-based training in these groups.

Table 4.15 End of compulsory schooling as determinants of participation

	Immediate	Later life
Regular school attender	4.03	1.79
Not regular attender	–	–
School attended at 16		
Comprehensive	4.71	
Grammar	11.94	
Secondary modern	2.99	
Private school	27.45	
Elementary/other	–	
First occupation		
Service class	13.00	1.89
Non-manual	2.65	3.37
Skilled manual	1.03	2.14
Part-skilled	0.80	1.47
Unpaid	–	–

Note
The coefficients are for categorical variables, and represent a change in odds compared to the final category. For example, someone who attended school regularly was over four times as likely to continue in education or training after school as someone who did not.

Table 4.16 shows the model coefficients for relevant variables relating to adult life and the present day. The figures show that despite all of the intervening variables in the model, there is still a key role for 'place' in explaining patterns of learning. However, here the patterns are reversed somewhat. People in the poorer areas of the Forest of Dean and Blaenau Gwent are now more likely to report later learning, once the social determinants of age, sex, family and so on have been dealt with. As already noted, adults with children were less likely to participate in later-life learning. Those who report sustained informal learning were also more likely to undertake formal learning in later life.

To what extent does use of ICT interrupt or reinforce existing patterns of participation in lifelong learning?

Having thoroughly examined the patterns of participation and non-participation in lifelong learning which were apparent in our data we can now move on to the last question posed at the beginning of the chapter – and one upon which this book is predicated. What role can ICTs be said to play in interrupting these patterns of lifelong learning? Can access to ICTs be said to be 'creating' adult learners? Probably the first and most obvious observation from our survey data about patterns of formal learning and access to and use of ICTs is that the two are strongly related. This is something on which most commentators are agreed (e.g. Sargant and Aldridge 2002; Kingston 2004a). The use of computers and the internet is strongly associated with social class, education, and location, indigeneity and birthplace (Gibson 2003).

The same relationship appears again in our new study (Table 4.17). Our indicators of access to and use of computers, and the internet, all show the same

Table 4.16 Adult life determinants of participation

	Immediate	Later life
Area of residence		
Forest of Dean		1.68
Bath and NE Somerset		1.08
Blaenau Gwent		2.15
Cardiff		–
Have children		0.23
No children		–
Informal learner		1.86
Not informal learner		–

Note
The coefficients are for categorical variables, and represent a change in odds compared to the final category. For example, someone with children is less than a quarter as likely to undertake education or training in later life as someone without a child.

Table 4.17 Patterns of participation by computer and internet use

Trajectory	Ever used a computer	Never used a computer	Used computer in last year	Not used computer in last year	Used internet in last year	Not used internet in last year
Non-participant	19	67	16	61	12	55
Transitional learner	20	13	22	13	24	13
Delayed learner	30	16	29	20	26	23
Lifelong learner	31	4	34	6	38	8

Note
Some columns may appear not to sum to 100 per cent due to rounding.

pattern – those who undertake formal learning are also those individuals who are most likely to use these ICTs.

However, we do not immediately draw the same conclusion as others have done. It is not necessarily the case that use of ICT leads to greater participation. The causal model could be the reverse of that, or both phenomena could have a common cause, or they could be iteratively related (Gorard 2002). In order to test this, we return to our logistic regression model. As we discussed earlier, the majority of the variance in patterns of participation that *can* be explained is explained by what we could have known about the respondents when they were born (e.g. year of birth, sex, father's occupation). Our model improves through the addition of what we could have known about the respondents when they reached school-leaving age (type of school attended, for example). Our model for participation in later life also improves through the addition of information about the respondents as adults (occupation and number of children, for example). But adding all of the generic variables about experience of ICT, access to ICT, and current use of ICT does not improve the model any further (see Table 4.18).

While this model is not in any way a definitive test, it does suggest that ICTs in themselves are not key determinants of adult participation in formal learning. The chief obstacles to participation reported by learners are anyway not the physical barriers of time and place that technology is supposed to overcome, but rather issues such as lack of interest or motivation (indeed, lack of interest was offered by 78 per cent of non-participants in La Valle and Blake's (2001) study as the reason for their non-engagement with learning). As one of our interviewees reasoned:

> there are plenty of places you can go and learn but I basically haven't got an interest in it ... I'm glad I don't have to learn anymore!
>
> (female, mid-aged, non-participant)

In summary the key determinants of extended initial education apparent in our survey data are parental occupation and education and respondent's age, place of birth, ethnicity and initial schooling. The key determinants of later participation are, in addition, area of residence, sex, occupation and having children. If these early-life determinants do create relatively stable learner identities or

Table 4.18 Percentage of patterns of participation correctly predicted at each life stage

	Variables at birth	Variables at end of schooling	Variables as adult	Level of ICT access
Immediate	79	84	84	84
Later life	71	71	77	77

Note
The use of information from adult lives cannot be said to 'predict' episodes of immediate post-compulsory initial education anyway.

trajectories, then it should be no surprise that supply-side 'solutions' such as ICT have little or no impact in later life.

Discussion

We have now conducted similar analyses on several datasets totalling 10,000 adults across the entire UK, and the same determinants appear each time (Gorard *et al.* 2003). Therefore, we are confident that these findings are robust. The key social determinants predicting lifelong participation in learning are time, place, sex, family and initial schooling. *When* respondents were born determines their relationship to changing opportunities for learning and social expectations. It is significant that respondents with similar social backgrounds from different birth cohorts exhibit different tendencies to participate in education and training. Time may be a composite proxy here for a variety of factors such as changes in local opportunities, economic development, the increasing formalisation of training, the antagonism between learning and work, and the changing social expectations of the role of women. Older respondents often reported quite radical changes of job or responsibility with no training provided at all.

Where respondents are born and brought up shapes their social expectations and access to specifically local opportunities to participate. Those who have lived in the most economically disadvantaged areas were least likely to participate in lifelong learning. This may be partly to do with the relative social capital of those in differing areas, or the changes in actual local opportunities to learn. However, those who have moved between regions are even more likely to participate than those living in the more advantaged localities. It may not be an exaggeration to say that those who are geographically mobile tend to be participants in adult education or training, while those who remain in one area, sometimes over several generations, tend to be non-participants. Men consistently reported more formal learning than women. Although the situation is changing, these changes are different for each sex. Women were still less likely to have participated in lifelong learning, but are now more likely to be 'transitional learners'. Extended initial education appears to now be relatively gender neutral, while later education or training is *increasingly* the preserve of males.

Parents' social class and educational experience are perhaps the most important determinants of participation in lifelong learning. Family background was influential in a number of ways, most obviously in material terms, but also in terms of what was understood to be the 'natural' form of participation. Experience of initial schooling appears to be crucial in shaping long-term orientations towards learning, and in providing qualifications necessary to access many forms of further and higher education. There are important 'age effects' here, however, relating especially to the reorganisation of secondary schooling in the maintained sector. For the older cohorts, ability-testing at the age of eleven via the 'eleven plus' examination was a clear watershed, and the ways in which the story played out from then on was clear from our interview data. Those who 'failed' at school often came to see post-school learning of all kinds as irrelevant to their needs and capacities. Hence, not only is participation in further, higher and continuing education not perceived to be a realistic possibility, but also work-based learning is viewed as unnecessary. There is thus a marked tendency to devalue formal training and to attribute effective performance in a job to 'common sense' and experience.

We have seen that over a third of the adult population has not engaged in any further learning at all and that those individuals participating in adult education are heavily patterned by 'pre-adult' social factors such as socio-economic status, year of birth and type of school attended. As we have already intimated, this confirms a long line of studies from the 1950s onwards which have provided compelling evidence that the determinants of participation (and non-participation) are long term and rooted in family, locality and history. If nothing else, our data confirm the first premise of the le@rning society thesis that there are a substantial number of individuals who currently do not participate in education for whom opportunities must be extended. But we have found that non-participation in adult learning is structured by factors which occur relatively early in life. This has profound implications for the applicability of ICTs in engineering a more inclusive learning society. Indeed, our analysis so far suggests that policies which simply make it easier for people to participate in the kinds of education and training which are already available (for example, removing 'barriers' to participation, such as costs, time and lack of child-care) will have only limited impacts. This includes approaches using new technologies.

It is clear from this analysis that we must examine our survey data in closer detail with regard to our respondents' use of technology. Although robust, our analysis has so far drawn on only a limited range of indicators of ICT access and use, and has concentrated primarily on learners' reported participation in formal learning. Having earlier gone to great lengths to stress the multi-faceted nature of adult learning as well as the many different forms of 'ICT', we need to revisit our survey data in more detailed terms of respondents' engagement with ICTs. How exactly are ICTs being used by respondents, if at all? Are ICTs just being used for non-educational purposes? Is there evidence of ICTs supporting informal rather than formal learning? The next chapter addresses these questions and explores further the relationship between ICTs and adults' learning.

Chapter 5

What do people use ICTs for?

Introduction

It appears from our analysis in Chapter 4 that ICT-based learning is having little impact on overall patterns of (non)participation in formal adult learning. The non-participation of adults in education remains a significant and deep-rooted trend, with or without ICT-based initiatives. Our conclusion so far is that whether or not an individual participates in learning appears to be a lifelong pattern, already presaged at school-leaving-age, and intrinsically related to long-term social, economic and educational factors. Crucially, access to technologies such as the computer or internet does not, in itself, seem to make people any more likely to participate in education and (re)engage with learning.

So far we have examined the survey data primarily in terms of education and formal learning. As such, we still know little about what learning activities ICTs are actually being used for amongst the adult population. In order to examine in more detail the apparently modest educational impact of new technologies, we now go on to consider *who* amongst the adult population in the UK is using ICTs and *what* learning they are using it for. Moreover, we examine who is *not* using ICTs for learning. With this in mind, the present chapter examines the following issues:

- Who among our household survey has access to various technologies with a capability of delivering learning experiences?
- What do adults report using ICTs for in their day-to-day lives?
- What is the level and nature of adults' use of ICTs for formal education?
- What is the level and nature of adults' use of ICTs for educative purposes in the workplace?
- What is the level and nature of adults' use of ICTs for educative purposes in the home and community?
- How are educative uses of ICTs patterned according to individuals' demographic characteristics?

Who among our household survey has access to various technologies with a capability of delivering learning experiences?

An important step towards understanding patterns of use of ICTs for learning is to gain a picture of patterns of *access*, especially the hierarchies of adults' access to different technologies (from actually owning a technology in the home through to shared access elsewhere). In line with other studies, our survey data showed that the most accessible technologies to adults were mass-market broadcast and communications technologies. The majority of respondents had access at home to fixed/landline telephones (90 per cent), terrestrial television (98 per cent), video recorders/players (89 per cent), CD players (78 per cent), mobile phones (75 per cent) and radios (95 per cent). As can be seen in Table 5.1, access to different types of computer technologies was lower – although 58 per cent of the sample had home access to a computer of some sort. The level of access to the internet was lower and predominantly through computers (42 per cent of the survey sample), rather than other internet-enabling technologies such as digital televisions (2 per cent) and mobile phones (2 per cent).

In terms of *where* adults were able to access computer-based technology, the most frequently cited locations were at home or the home of a relative (Table 5.2). The next most common locations were the home of friends, places of work and libraries. Just over one third of respondents (34 per cent) reported having access to some form of public ICT site – with most of these respondents citing libraries (28 per cent), commercial pay-per-use sites (14 per cent) and local educational institutions (10 per cent) as offering potential access to computers if they needed it. Only 5 per cent of respondents cited having access to computers in community centre sites. Public access sites were substantially less likely to be cited than the homes of friends and relations.

As we argued in Chapter 2, there is a need for social research to develop more detailed understandings of how access to new technologies is patterned. With this in mind the models of access proposed by Wilhelm (2000) and Murdock (2002) which seek to identify the degrees (or layers) of connectivity/marginality to ICTs are adopted. At the centre of this hierarchical model are *core access* individuals who have ready access to computers at home and enjoy access to advice and support that enables them to operate more effectively and to continually extend their range of uses (Murdock 2002). A second category occupied by those individuals who have access at home but are limited by ageing equipment and limited support is *peripheral home access*. Not in Murdock's original description (but arising from our data) are those individuals who lack access to computers in their home but do have access through family and friends as well as terminals in public locations or at work alongside access to limited support (*peripheral family access*). Yet another group are those individuals whose sole access is through shared terminals in public locations or at work, where their use is heavily constrained by the demands of other users and limited support (*peripheral public access*). The

Table 5.1 Adults' access to technologies.

Information and communication technology	Own/access at home	No home access but access from family/friends	Access only at work	Access only elsewhere	No access
Laptop computer	9	2	4	2	83
Palmtop computer	4	1	1	1	93
Desktop computer <5 years old	39	9	6	2	44
Desktop computer 5+ years old	12	3	2	1	82
Computer printer	45	8	5	1	41
Computer scanner	29	7	6	2	56
Digital camera	15	5	3	1	76
Digital television	33	7	0	1	59
Video recorder/player	89	0	0	0	11

Notes
Data are percentage of respondents ($n = 1,001$). Categories of access are mutually exclusive. Summed data may not add up to 100 per cent due to rounding up and rounding down of decimal places.

Table 5.2 Adults' perceived access to computers

Site of access	Percentage
Your home	58
A relative's home	47
A friend's home	29
Your workplace/place of study	29
A library	28
A private 'pay-per-use' site (e.g. internet café)	14
A local school/college/university (non-students)	10
A community centre/site	5
A museum/science centre	3

Note
Data are percentage of respondents ($n = 1,001$).

most peripheral are those individuals who have no ready access to computers or support at all (excluded).

Using the access and ICT support data from our survey, it is possible to assign our respondents to one of these five groups. Access to computers was calculated from the data summarised in Tables 5.1 and 5.2, whilst access to computer support was calculated in terms of respondents' reported sources of support. Here, 'ready access to a range of ICT support' was defined as being able to access two or more sources of support in answer to the question 'who of the following, if any, could you go to for help/advice if you wanted to use a computer?' Respondents citing only one source were classed as having 'limited support'. This analysis reveals a more delineated picture of adults' ICT access than is suggested in the existing literature (Table 5.3). Only 8 per cent of our sample can be classed as being absolutely excluded from computer access. Conversely 50 per cent reported having ready access to a computer in a home setting (albeit only 21 per cent with up-to-date resources and a range of support). Therefore, 42 per cent of adults are reliant on some form of outside-home peripheral access. As was suggested above, this peripheral access is supplied for most people through the extended family rather than at public or community sites. In terms of differences between 'core' to 'excluded' categories of access, some variations were apparent according to respondents' age, illness/long-term disability, educational background and socio-economic status – although not in the case of sex. To some extent these replicate the determinants of participation in formal learning.

What do adults report using ICTs for in their day-to-day lives?

As we also observed in Chapter 2, having access to a technology cannot be equated with use – and 'use' cannot be necessarily equated with 'meaningful' use. Many people in previous surveys have reported having access to the internet but not using it, for a variety of reasons such as lack of privacy, and the intrusion of unwanted material (Mathieson 2003). With this in mind, we can now turn our attention towards how respondents reported making use of their access to computers and, in particular, explore how computers were being used for educative purposes. It is worthwhile first providing a general context for patterns of educative use of computers by considering briefly the overall patterns of any computer use reported in the household survey. Our initial observation in this respect is that computer use is by no means a ubiquitous activity within the adult population. Although only 8 per cent of the survey sample could be classed as being totally 'excluded' from computers, 48 per cent of the sample reported not having used a computer during the previous twelve months. Indeed the use of computers remained a minority activity in the home when compared with the use of other technologies such as television, video/DVD, radio, hi-fi and the mobile phone. Watching television and listening to the radio were the most popular technology uses among the sample, with 93 per cent watching television frequently (i.e. 'very' or 'fairly often'), and 81 per cent listening frequently to the radio.

Within the 52 per cent of the sample who had used a computer, word-processing was the most popular activity, followed by 'fiddling around on the computer', file and memory organisation, and learning from computer software (Table 5.4). In terms of use of the internet, sending and receiving emails was the most prevalent internet-based activity, alongside searching for information on goods and services. The relatively low levels of use of ICT for learning purposes needs to be seen within the context that the majority of respondents displayed a limited 'repertoire' of uses of computers and the internet. Of the 22 listed applications the mean number used on a 'frequent' (i.e. 'very' or 'fairly often') basis was six applications (standard deviation (s.d.) = 4).

Table 5.3 Level of access to computers by social, health and education characteristics

Hierarchical level or category of access						
Social or health characteristic	Core access	Peripheral home access	Peripheral family access	Peripheral public access	Excluded	Sample size
Sex						
Male	21	33	29	11	7	405
Female	21	26	31	13	9	596
Age group (years)						
21–40	26	35	26	9	4	330
41–60	31	35	22	9	3	319
61 or more	7	17	41	19	17	352
Marital status						
Single/separated/widowed	13	17	40	19	11	355
Married/living with long-term partner	26	35	25	9	6	625
Health status						
No long-term illness/disability	23	31	27	11	7	761
Long-term illness/disability	14	20	38	17	11	229
Education						
Continued after 16 years old	28	41	17	6	7	384
Completed education at or before 16 years of age	17	21	38	16	9	617
Socio-economic status						
Service	29	47	11	7	6	83
Skilled non-manual	28	32	21	13	6	300
Skilled manual	22	43	24	7	5	87
Part-skilled	16	20	40	13	11	418
Other	14	28	34	17	7	113
Total	21	29	30	12	8	1,001

Note: Data are percentage of respondents (n = 1,001). Summed data may not add up to 100 per cent due to rounding.

Table 5.4 Use of computers and the internet in the last 12 months

Activity or use	Very often	Fairly often	Rarely	Never
Writing and editing letters, reports and other documents	27	11	8	54
Send/read emails (via computer or digital TV)	26	7	5	62
Look for products and services/gathering product information online	14	15	7	65
Fiddling around on a computer/explore different bits of the computer to develop your own knowledge	14	14	8	66
Look for information related to work/business/study on the world wide web	15	10	5	69
Organising the computer's files/memory	11	11	9	69
Learn something when using a computer program (e.g. from a CD-ROM, encyclopaedia or database)	9	13	8	70
Buy goods and services online	4	9	9	78
Browse/surf the world wide web for no specific purpose	4	7	10	79
Playing games	4	5	10	81
Creating and manipulating images (e.g. photographs)	5	6	7	81
Listening to music on a computer (CDs, MP3s)	3	7	7	82
Online banking/management of personal finances	7	5	4	84
Download software, music, films or images from the internet	4	5	7	85
Programming the computer	3	4	6	87
Use internet newsgroups, bulletin boards, chat rooms or instant messages	2	4	4	91
Watching DVDs/videos on a computer	1	3	5	91
Making music with a computer	2	3	4	91
Participate in educational courses/lessons on the world wide web	2	4	4	91
Making/maintaining your own website product information	3	2	2	94
Making films or animations on a computer	2	2	3	94
Use adult entertainment on the world wide web	0	1	3	96

Note
The reported data are percentage of respondents ($n = 1,001$). We were surprised at the relatively low level of reported use of the internet for adult entertainment, given its prevalence in other estimates of online activity. Of course, this could be an artefact of the methods used.

Moving our attention from the nature of use to location of use, only 11 per cent of respondents from the survey reported making use of computers in some form of public ICT site during the past twelve months – as opposed to 44 per cent making use of ICT at home and 32 per cent making use of ICT in the workplace (see Table 5.5). Mirroring the patterns of perceived access presented earlier, the greatest number of those respondents making use of ICT in public sites had done so in libraries and local educational institutions (4 per cent and 5 per cent respectively). Only 2 per cent of respondents were making use of ICT in community sites and 3 per cent of respondents in commercial 'pay-per-use' sites. This patterning is not unique. If we compare these figures, for example, to those from La Valle and Blake (2001) we can see that 83 per cent of those who had used a computer did so in their own home, 45 per cent in their place of work, 5 per cent in a library, and less than 1 per cent in a community centre or job centre.

What is the level and nature of adults' use of ICTs for formal education?

Within these general patterns of use, how then are computers and the internet being used for education and learning purposes? If we consider first the use of computers for formal education and learning then we find that 21 per cent of our respondents reported having used a computer during post-compulsory educational episodes. As can be seen in Table 5.6, computers were used most often as an adjunct to non-technology methods of teaching and learning. Belying the rhetoric of 'virtual learning', very few respondents had been fully taught via ICT and fewer still had used ICT to find out about learning opportunities. This is confirmed by other sources. Gorard (2003), for example, reports that *no one* in the annual NIACE survey of adult learning in Wales had found out about their latest episode of learning via the learndirect web portal or telephone helpline. More common

Table 5.5 Frequency of use of PCs/computers in different locations over the past 12 months

Location	Very often	Fairly often	Rarely	Never
Your home	25	12	7	56
A relative's home	1	3	10	86
A friend's home	1	2	7	90
Your workplace/place of study	25	5	2	68
A museum/science centre	0	0	3	97
A community centre/site	0	0	1	98
A private 'pay-per-use' site (e.g. internetcafé)	0	1	2	97
A local school/college/university (non-students)	1	1	3	95
A library	0	1	3	96

Note

Data are percentage of respondents (*n* = 1,001). Percentages less than 0.5 per cent appear as 0.

Table 5.6 Use of ICT in post-compulsory formal education

	n	%
Found out about study using ICT	20	2
Assessment involved ICT	89	9
Used computers to research information	126	13
Used computers to complete assignments	145	14
Partially taught via ICT	84	8
Taught to use computers/software	103	10
Fully taught via ICT	28	3

Note
Data are percentage of respondents (*n* = 1,001).

was the use of ICT to research information and produce assignment materials for traditional face-to-face courses – alongside actually being taught to use ICT as part of a wider formal education programme. Indeed, reflecting the rapid growth of computer skills courses in the post-compulsory and adult education sectors, 10 per cent of our respondents reported having taken elementary courses in ICT not leading to a recognised NVQ-level qualification.

What is the level and nature of adults' use of ICTs for educative purposes in the workplace?

Around 40 per cent of our respondents reported having, or having once had, a job which involved using ICT in some form. Just over one-quarter of these (11 per cent of the overall sample) reported having been trained at work using ICT in some form. As can be seen in Table 5.7, this most frequently (but not exclusively) involved being taught to use computers or specific software packages, with using ICT to research information and complete assignments also being relatively common uses of ICT in work-based training. Again, being trained fully via ICT was cited only by a minority of respondents. These relatively low levels of formal use of ICT in work-based training are explained later, where the reliance on informal learning is being highlighted by respondents, and in particular the process

Table 5.7 Use of ICT for work-based training

	n	%
Found out about study using ICT	16	2
Assessment involved ICT	54	5
Used computers to research information	58	6
Used computers to complete assignments	70	7
Partially taught via ICT	47	5
Learnt to use computers/software	73	7
Fully taught via ICT	20	2

Note
Total *n* = 1,001.

of learning to use a computer for a job via an 'informal apprenticeship'. This is explored more fully in Chapters 7 and 9.

What is the level and nature of adults' use of ICTs for educative purposes in the home and community?

In terms of using ICTs for learning at home (and to a lesser extent in community sites), computers and the internet tended to be used, if at all, for informal rather than formal learning. For example, only 6 per cent of the overall sample (59 people) reported using the internet more than 'rarely' for participating in educational courses/lessons via the world wide web, whereas 30 per cent reported 'learning something' from a computer program whilst using it (see Table 5.8). Similarly, informally seeking information relating to work, business or study via the internet was a more frequently mentioned learning activity, as was learning about using the computer itself through 'fiddling around'.

Continuing this theme of informal learning, 26 per cent of the survey sample reported having a sustained hobby or leisure pursuit which had involved them having to learn something. These ranged from practical pursuits (including DIY and household maintenance) to art, music, sport and using computers as a pursuit in itself. As can be seen in Table 5.9, just over a third of these learners (n = 85) had used ICT in some way to support this 'informal' learning – mainly for researching information about the hobby or leisure activity.

Table 5.8 Frequency of use of PCs/computers for educative purposes at home

	Very often	Fairly often	Rarely	Never
Participate in educational courses/lessons on the www	2	4	4	91
Learn something via CAL package, CD-ROM, encyclopaedia, database	9	13	8	70
Look for information related to work/ business/study	15	10	5	69
Fiddling around with computer	14	14	8	66

Note: Data are percentage of respondents (n = 1,001).

Table 5.9 Use of ICT for sustained informal learning at home

	n	%
Used computers to research information	66	7
Learnt to use computers/software	26	3
Assessment involved ICT	17	2
Partially learnt via ICT	15	2
Found out about area of study using ICT	8	1
Fully learnt via ICT	3	0

Note: Data are percentage of respondents (n = 1,001).

This use of ICT for supporting informal learning is worthy of further attention, especially as informal learning was not encompassed by the four lifelong learning 'trajectories' discussed in Chapter 4. Therefore the suggestion in Table 5.10 that there is a greater tendency for informal learners to use a computer (and the pattern is the same for access and use of the internet), is initially an encouraging one. It could be that although technology cannot overcome the impact of long-term socio-economic determinants on patterns of adult participation in formal learning, it can act as an equaliser for those who wish to learn for themselves anytime, anywhere, anyhow.

In fact, our survey data do not support this interpretation either. Table 5.11 shows that the link between informal learning, formal learning and ICT-use is also strong. The first column shows that those who learn formally are also more likely to learn informally (only 17 per cent of non-participants reported informal learning). The second column shows that those who learn formally are also more likely to involve ICT in their informal learning (20 per cent of lifelong learners reported informal learning using ICT).

Nevertheless, there were 67 individuals in our survey who report no formal education or training since reaching school-leaving age but who are sustained informal learners, of whom 24 have used ICT in their learning in a variety of ways (e.g. using computers to research information, finding out about their area of study using the internet or even fully learning via ICT). Neither of these two groups of informal learners differed substantially from the wider sample in terms of age, sex, household composition (e.g. marital status, children, etc.). The only discernible feature that sets them apart is that 58 per cent ($n = 14$) of the ICT-based sustained informal learners and 55 per cent ($n = 37$) of the overall group of

Table 5.10 Patterns of computer usage by sustained informal learning

Any informal learning?	Used a computer in the past 12 months	Not used a computer in the past 12 months
No	50	50
Yes	75	25

Note
Data are percentage of respondents ($n = 1,001$).

Table 5.11 Patterns of (computer-based) informal learning by later post-compulsory participation

Trajectory	Informal learning	Informal involving ICT
Non-participant	17	6
Transitional learner	24	10
Delayed learner	29	10
Lifelong learner	38	20

Note
Data are percentage of respondents ($n = 1,001$).

67 are from the 'partly skilled' socio-economic group (a much higher proportion than in the sample overall).

Who is, and who is not, using ICT for educative purposes?

Returning to our more general question of how educative uses of ICTs differed according to individuals' demographic characteristics we can see that all of the uses of technology for learning within our survey sample were stratified according to a variety of demographic variables. Prominent differences in use were evident by socio-economic status, age group, marital status, area of residence and educational background (Table 5.12). Less pronounced differences in use were also apparent by long-term illness/ disability and household composition. Although the differences were slight, men were more likely to have used ICT for informal learning, whereas women were more likely to have used ICT for formal learning. The nature and level of use of computers for both formal or informal learning differed according to level of access to ICT – with a clear divide in favour of those individuals with access at home compared to those relying on family or public access. Finally if we examine respondents' histories of lifelong learning in terms of their 'trajectories' of lifelong education we can see that, in general, use of ICT for educative applications increases with levels of educational engagement (i.e. those individuals involved in more sustained levels of learning also tend to be more likely to use ICT for learning). There are, however, some interesting exceptions. For example, transitional learners (those who reported at least one episode of immediate post-compulsory education or training but nothing subsequently) were more likely than any other group to have used the world wide web to participate in formal online courses and/or lessons. Of course, these differences will also be related to the age differences between the trajectories.

Overall, 51 per cent of the sample reported using a computer for one or more of the 'educative purposes' in Tables 5.12 and 5.13 – just 1 per cent less than the proportion of the sample who reported using a computer during the past twelve months. These included 15 per cent of the non-participants in formal learning, 55 per cent of the transitional learners, 64 per cent of the delayed learners, and 86 per cent of the lifelong learners. Therefore, learning via a computer whether informally or not appears to be as stratified as learning in institutions or via any other medium. As we demonstrated in Chapter 4, we were able to 'predict' the learning trajectory of an individual just from their background characteristics. Can we predict learning experiences via a computer in the same way?

To answer this question we can return to the multivariate analysis technique utilised in Chapter 4. For the purposes of the data relating to respondents' ICT use, we collapsed all of the reported types of learning experience using computers or the internet (as described in Tables 5.12 and 5.13) into a binary variable representing whether each individual had used a computer or the internet for any 'educative purposes' or not. We then used this as the dependent variable in a

Table 5.12 Usage of computers for formal educative purposes by personal characteristics

	Taken formal course in ICT (elementary)	Used ICT in post-compulsory education episodes	Formally learnt/ trained to use ICT in work	Participate in educational courses/lessons on the www	Sample size
Sex					
Male	9	19	12	7	405
Female	12	23	12	5	596
Age group (years)					
21–40	11	42	16	9	330
41–60	14	20	14	7	319
61 or more	7	3	5	1	352
Socio-economic status					
Service	10	29	22	15	83
Skilled non-manual	16	31	21	5	300
Skilled manual	6	14	10	6	87
Part-skilled	8	12	5	4	418
Other	12	27	5	9	113
Marital status					
Single/separated/widowed	9	17	7	4	355
Married/living with long-term partner	12	24	15	7	625
Health status					
No long-term illness/disability	10	24	13	7	229
Long-term illness/disability	12	12	7	4	761
Education					
Continued after 16 years	16	40	18	10	384
Completed education at or before 16 years of age	8	9	8	3	617

Area of residence					
Bath/NE Somerset	12	23	8	7	253
Blaenau Gwent	10	13	11	4	248
Cardiff	7	24	14	10	251
Forest of Dean	14	23	14	3	249
Household composition					
Single adult aged 16–59	9	38	14	9	124
Small family	12	31	17	8	349
Large family	16	27	14	10	163
Large adult household	0	6	12	12	17
Adult aged 60 and over	6	2	1	1	192
2 adults, 1 or both aged 60 and over	9	5	7	1	156
Hierarchical level of access to ICT					
Core access	15	41	21	10	210
Peripheral home access	13	33	16	12	286
Peripheral family access	8	7	5	0	299
Peripheral public access	6	6	8	2	124
Excluded	6	6	1	0	82
Lifelong learning trajectory					
Non-participant	0	0	0	2	371
Transitional learner	0	28	1	11	175
Delayed learner	19	23	19	6	246
Lifelong learner	30	51	33	9	209
Total	11	21	12	6	1,001

Note
Data are percentage of respondents in sample ($n = 1,001$).

Table 5.13 Usage of computers for informal educative purposes by personal characteristics

	Learn something via CAL package, CD-ROM, encyclopaedia, database	Look for information related to work/ business/study	Used ICT for sustained informal learning	Fiddling around with computer	Sample size
Sex					
Male	23	27	10	32	405
Female	21	24	7	24	596
Age group (years)					
21–40	32	39	8	40	330
41–60	25	32	14	32	319
61 or more	9	7	3	10	352
Socio-economic status					
Service	36	48	15	36	83
Skilled non-manual	25	37	11	33	300
Skilled manual	28	35	13	40	87
Part-skilled	16	13	6	17	418
Other	20	19	5	33	113
Marital status					
Single/separated/widowed	16	17	7	20	355
Married/living with long-term partner	25	31	9	31	625
Health status					
No long-term illness/disability	24	29	9	29	229
Long-term illness/disability	15	14	8	21	761
Education					
Continued after 16 years	32	42	13	38	384
Completed education at or before 16 years of age	15	15	6	20	617

Area of residence					
Bath/NE Somerset	17	23	10	18	253
Blaenau Gwent	19	16	2	25	248
Cardiff	27	36	9	33	251
Forest of Dean	25	28	12	32	249
Household composition					
Single adult aged 16–59	32	35	14	40	124
Small family	27	35	12	33	349
Large family	31	37	10	41	163
Large adult household	12	6	0	24	17
Adult aged 60 and over	4	3	3	3	192
2 adults, 1 or both aged 60 and over	13	12	4	15	156
Hierarchical level of access to ICT					
Core access	43	51	19	52	210
Peripheral home access	37	43	13	47	286
Peripheral family access	3	4	2	5	299
Peripheral public access	7	11	3	8	124
Excluded	0	0	0	0	82
Lifelong learning trajectory					
Non-participant	7	7	4	9	371
Transitional learner	26	32	7	32	175
Delayed learner	27	28	9	37	246
Lifelong learner	37	51	17	44	209
Total	22	26	9	27	1,001

Note
Data are percentage of respondents in sample ($n = 1,001$).

logistic regression analysis using all of the personal characteristic variables in Table 5.12 as predictor variables. As we outlined in Chapter 3, logistic regression relies on far fewer assumptions about the data than alternatives such as linear regression or discriminant analysis, and makes the use of categorical predictor variables considerably easier. We created the model of ICT usage for learning (or not) in two stages, using backwards stepwise selection of the variables for each stage. As with our earlier analysis of the participation data, in the first stage we added the variables that could be known about the individual from birth (age, sex, ethnicity and so on). In the second stage we added variables about their current life (household composition, health status and so on).

Using only those background variables that we could have known about each individual since birth, we can predict their later use of ICT for learning with 69 per cent accuracy (or put another way we can improve any guess due only to chance by 31 per cent). In producing this model, factors such as first language, family religion and ethnicity of each individual were found to be irrelevant if other factors were taken into account at the same time. The only background variables of substantive relevance were *age* (ICT use for learning declined by around 0.94 for each year of age), and *sex* (men were nearly 1.4 times as likely to report using ICT for learning).

Using background variables known about each individual now, we can improve our prediction about the use of ICTs for learning to 82 per cent accuracy (or put another way we can improve any guess due only to chance by 60 per cent). In producing this model, ill-health, marital status, number of children, household composition, and the geographical mobility of each individual were found to be irrelevant once other factors had been taken into account. The background variables of substantive relevance for this second stage of the model were:

- *Continuing with education or training at age 16* – those who did so were 2.6 times as likely to learn using ICTs, whether informally or not, as adults aged 21 or more;
- *Occupational class* – those in the professional/service class were 2.4 times as likely to learn via ICTs as the unskilled or part-skilled;
- *Area of residence* – Those living in the remote Forest of Dean were 1.4 times as likely to learn via ICTs as those in urban Cardiff, while those in Bath and North East Somerset (0.81) and Blaenau Gwent (0.69) were less likely to learn via ICTs than those in Cardiff.

Although having less of a bearing once occupational class and educational background have been taken into account, the role of geographical location is especially interesting. As previous authors have shown, the general consumption and take-up of technologies is not uniform – there are local contingencies and specificities that work alongside global influences in local consumption processes of ICTs (Williams 1999; Miller 1994; Murdock *et al.* 1996). Similarly from our data, it appears that the take-up of technologies for learning and education differs significantly between localities – even when controlling for socio-economic status and other local characteristics. Of course, location may be acting as a proxy for other variables not included in our regression model but this impact of geographical location merits further exploration in later chapters.

ICT and learning – the views of participants

We conclude this chapter by illustrating some of these patterns through an initial investigation of the in-depth interview data collected from 100 of the survey respondents. We can use these data to elaborate on the very low numbers of respondents using ICT as a vehicle to engage in formal learning. The various reasons underlying this reticence can be seen in the example of one of our interviewees who, in theory, would be a prime candidate for ICT-based learning. She ran her own web-development company, used an extensive range of ICTs on a daily basis (laptop computers, personal digital assistant, internet-enabled mobile phone, MP3 player) and was a self-confessed 'learning addict' having completed a variety of adult and post-compulsory courses since leaving school. Yet even for this high technology-using, lifelong learner the prospect of ICT-based formal learning was not appealing, firstly for social reasons:

> Interviewer: But have you been tempted by all the online courses you can take, never actually having to leave the comfort of your front room?

> I'll tell you what puts me off those – I've had a scan through the learndirect courses – and it's the feeling that they're trying to teach basic skills without teacher interaction, and I personally like classroom interaction. And I don't think you can get the same buzz doing it online. I chat [on the internet] quite often to friends in the States. In chat rooms the difficulty is that it becomes very disjointed and you lose threads very easily and you lose the interaction that you get when you're face to face. And I think that's the disadvantage of it … if I wanted to learn Maths or something I think it would be great. But I think if you were learning something that required a bit more interaction, I would treat it with a bit of distrust.
>
> (female, 38 years)

… but also through her perceptions of formal online learning provision in the UK:

> I get the feeling that learndirect is aimed at people who don't have … who aren't interested in learning, who are lacking in the paper qualifications that they need in order to be able to take them back out into the jobs market. And so, for me, it's an electronic YTS scheme [Youth Training Scheme].
>
> (female, 38 years)

Crucially, as this recently retired camera enthusiast who had also developed an interest in the internet explained, the opportunity to be able to formally learn online is of little use if there is no motivation to do so:

Interviewer: Would you consider doing a formal photography course on the internet?

Yeah, there are camera courses. I've thought about it, but I've probably got to the stage now that I don't want to be bothered. I think I've learnt enough, but I pick most things up. I can sit down and read something on the computer and I'd have the gist of how to do the job. My lad phones me up and says 'oh, are you going to do so and so, so and so?' Soon as he's told me, I can do it.

Interviewer: So you're still learning stuff all the time?

Yeah, well we all are, aren't we? ... I've thought about [doing an online course] but I never bothered. What can I say? I've really got to say, 'right, I want to do that thing', and I'll do it, but I just don't feel motivated at the moment to say that.

(male, 63 years)

These attitudes towards ICT-based formal learning permeated our interviews (and will be elaborated on in subsequent chapters). From our interview and survey data it was apparent that if ICT was stimulating participation in formal or institutionally-based non-formal learning then this was largely in terms of learning about using the technology itself – i.e. computer skills courses such as 'computers for the terrified', BBC Webwise and Computers Don't Byte courses, RSA CLAIT and the European Computer Driving Licence. But learning *about* ICT is very different from learning *through* ICT – a caveat often overlooked in the 'headline' figures behind government initiatives such as Ufl and learndirect.

Our survey data suggested that if ICT was being used to facilitate learning then it was predominantly in terms of informal learning. In our interviews we also found evidence of a range of informal learning taking place which involved ICTs. Of course, as with the formal learning, much of this involved informal learning to use ICT itself either 'on-the-job' at work or at home. One conclusion that emerges from our data is that the computer is a very self-referential learning tool. In other words, if ICT is being used by adults for formal or informal learning then this is often with regard to the computer itself. Learning about the computer was a common theme throughout our survey and interview data, be it in the form of taking ICT skills and computer literacy courses in the adult education sector or informally 'messing around' and 'playing' with the computer in order to learn how it works. In this respect ICTs have altered the subject but not necessarily the intensity of adult learning.

That said we did find evidence of informal learning through computers – some of which could be classified as sustained and part of a large-scale learning 'project' which people were engaging in. For some authors this form of 'intentional' informal learning is predominantly workplace focused – 'activities initiated by people in work settings that result in the development of their professional knowledge and skills' (Lonham 2000: 84). Yet, one of the interesting patterns to emerge from our study was the role of the computer in allowing this work-based informal learning to permeate beyond the workplace and into the home settings. We also found evidence of similar activities linked to the household, leisure and general interest domains – involving people using the internet and CD-ROMs to learn about subjects as far-ranging as car maintenance and chicken-rearing. However, it should be noted that these ICT-based episodes of intentional informal learning were usually part of an ongoing sustained project that also involved learning via books, courses and social contacts.

Also evident from our interviews, but more difficult to pin down, was the range of 'unintentional' or 'incidental' informal learning which was taking place via ICT – learning which was smaller scale but incremental in its effects. It was clear that for some people, ICTs such as the internet and computers were contributing in this way to forms of passive education. This is illustrated, for example, by our retired camera enthusiast interviewee. Throughout the course of his interview he revealed a range of subjects which he could be considered to be extremely knowledgeable about – from the transmission of television and radio signals and astrophysics to genealogy and medicine. Although he had claimed at the beginning of the interview not to use a computer extensively, as the interview progressed it became clear that much of this knowledge was the product of internet-based learning (he had also thoroughly researched our university department and research project website before the interview!):

> I should think I'm on [the internet] every day for something ... and if I want some information, I'll stay on there, you know ... All that information is on there if you keep looking for it. Ask Jeeves, 'could you please state where all television transmitters are'. And then it will give you a list of websites to go into ... I've got a program on there that will tell me where every speed camera is in England! So if I'm going somewhere long distance, I'll tap out the journey. There's a link to everything practically. I find I can get – well you can get in anywhere round the world. You can actually get into space satellites now.
>
> *Interviewer: Yes. I remember when you could look down the Hubble telescope.*
>
> You can still get in it now! I've got that on there [points to the computer] – I can just go into what that's looking at at the moment.
>
> (male, 63 years)

Yet, like most of our respondents who were informal learners there was little evidence that ICTs had created a new-found desire for learning – rather that it was building upon previous learning behaviours and disposition:

Interviewer: How do you pick up a lot of information and knowledge?

Probably 'cause I sit here for an hour and just start clicking through. Nothing general, just being nosy!

Interviewer: So it's something you were doing before you got the computer? Are you picking up more information now?

I think it's a lot of information that's at the back of my mind, you know. I've picked it up before and [the internet] is just jogging it up more. Just bringing it all out. I'm making up for lost time.

Interviewer: Do you ever touch your books now?

Yeah, oh yeah … Quite often, I'll pull a book out if I'm after some information. I go to the book first and if I can't find it in the book, then I go on the internet. 'Cause I think, why pay twice? I've got it in the book, if it's there.
(male, 63 years)

Thus, whilst ICTs could be not considered to be effective in facilitating formal learning amongst our respondents, there were strong suggestions that new technologies were aiding and augmenting some people's informal learning activities. Moreover, informal learning via the internet or computers was relatively appealing and effective for adults who were not successful formal learners, as these two quotations from a middle-aged mother and housewife illustrate:

There's not a lot of things I can't turn my hand to, that I will have a go at. But in learning aspects – I'd rather learn by watching and having a go myself. It's learning by experience, I suppose … you learn through life really, through everyday things. I've got nothing academic.
[…]
If I do go on [the computer] I find that I haven't forgotten things. I find it interesting. When I was in school – or even now – I'm not one for reading books. If I read something I get impatient. I start to read a book and I want to look at the end to see what the ending is and it doesn't seem to sink in. I forget things easy. But on a computer, because of the things you've learnt, I suppose it's like watching the television. If you watch a programme, the next day you can tell a friend exactly what, everything that happened on that programme. But if you've read it in a book, you forget. I don't know what it is. I suppose on the computer I've remembered everything [that I have learnt].
(female, 46 years)

Discussion

In providing a more detailed description of adults' (non)use of ICTs this chapter reaches some rather bleak conclusions in terms of the le@rning society thesis. First and foremost, it seems that when computers are being used by adults, then education and learning are minority activities. Moreover, any educative use appears to be patterned by a number of entrenched social factors. This continued stratification is important as it reiterates the point that access to ICT is unlikely, in itself, to make people any more likely to participate in education and (re)engage with learning. In this chapter we have shown how access to new technologies continues to be largely patterned according to long-term pre-existing social, economic and educational factors. Thus, like educational qualifications, ICT access is a proxy for the other, more complex, social and economic factors that pre-date it rather than a direct contributory factor in itself. Crucially, this stratification continues across the range of formal and informal learning activities which adults reported using computers for. We conclude that, at best, ICTs increase educational activity amongst those who were already learners rather than widening participation to those who had previously not taken part in formal or informal learning.

As we saw from our survey and interview data, whilst learning and education are not common occurrences within general patterns of ICT use, any educative use, more often than not, appears to be 'indirect' and informal rather than formally provided. Although there was little suggestion of ICT 'creating' new informal learners (i.e. it would seem that computers and the internet were mainly of use to those who would be informally learning anyway), the importance of informal learning in people's educative use of ICT should not be underestimated. Of course, the relationships between ICT and informal and formal learning are difficult to capture in a 'one-off' research activity such as a survey. So far we have only been able to touch upon some of the characteristics and themes of what emerges from our data as an important area of adult learning. Nevertheless, like all learning, participation in ICT-based formal and informal learning appears to hinge not on accessibility, cost or time constraints (although all are important as a prerequisite) but on underlying issues such as compulsion, motivation and disposition. This lends weight to the arguments outlined in Chapter 2 that there are more fundamental problems for participation in education than purely technical and infrastructural issues. Instead, as with all adult learning, the key factor underlying relative success of ICT-based learning would seem to be the motivation and sometimes learning discipline of the learners:

> Students must maintain persistence, enthusiasm, personal commitment and a clear focus in order to succeed in a distance learning situation. Self-direction, a passion for learning, and strong individual responsibility are important influences on achievement.
>
> (Frank *et al.* 2003: 59)

In this way ICTs would appear to be more motivating for informal learning than formal learning – or at least we can conclude that ICTs are proving more appealing to motivated informal learners than to motivated formal learners. How ICTs work in this way for learners, and how ICTs can be encouraged to work in this way for current non-learners, is explored further in subsequent chapters. It could well be, as Cullen *et al.* (2002: i) argue, that 'informal learning *is* widening participation'. If so, there are plenty of issues where we need to develop a better understanding. For example, how ICTs such as computers and digital television complement or usurp 'traditional' sources of informal learning such as books, analogue television and social networks; how the increased mobility of technologies such as computers and the internet may be facilitating the 'spread' of informal work-related learning into the home and people's leisure time; and how the outcomes of ICT-based informal learning differ to non-ICT based methods. We, therefore, need to develop our understanding of what ICTs can and cannot achieve in informal learning if we are to harness their undoubted educational potential.

Conclusion

From the initial engagement with our interview data in Chapters 4 and 5 we have glimpsed a number of pertinent issues worthy of further investigation. These are largely issues which cannot be addressed from survey data alone. These include (amongst others): learner motivation and interest, the relative unattractiveness of formal education in relation to informal and self-education, and the role of social contexts such as the home and workplace on ICT use. With this in mind the next four chapters describe our more in-depth data collected from interviews and ethnographic case studies. These data are examined in relation to the three main social contexts of technology-use evident in the survey data – the home, the workplace and public ICT sites.

Chapter 6

Adult learning with ICTs in the home

Introduction

Over the next three chapters we examine how adults engage with ICTs and ICT-based learning in the contexts of the home, the workplace and public sites such as libraries, museums, colleges and community centres. A fourth chapter then examines the specific case of how people learn to use the computers and the internet, thus providing a case study of how domestic, work and community contexts interact with each other in shaping (and being shaped by) adults' ICT use and learning. Through a careful examination of our interview and ethnographic case-study data the four chapters highlight a plethora of important but easily overlooked issues which lie at the heart of the le@rning society debate. These complexities faced by the apparently straightforward addition of ICT to adult learning are perhaps nowhere more evident than in the social context of the home.

The home as a site of the le@rning society

The home is now considered to be the key social context for many adults' access to and engagements with ICT. Most homes in developed countries are rich in technological provision as computers and the internet have become an integral part of the domestic leisure and entertainment activities. As Rob Shields reasons:

> Although less obvious than the workplace in terms of machinery, the domestic sphere is penetrated by communications technologies such as the telephone and television and stuffed with toys and small devices in which chips are embedded, allowing digital forms of virtuality to spread quickly into children's and families' lives in North America and Europe. Even in countries which lagged behind the trend-setting economies, the [domestic] take-up of the computer was one of the most rapid disseminations of new technologies on record.
>
> (Shields 2003: 94)

Aside from actual use, the home is also a key site within which adults think about and 'socially construct' their notions and understandings of technology. For many adults the home rather than the workplace is the main site where they get to use and explore technologies like the computer on their own terms – enjoying a control over the circumstances under which they engage with ICTs not available in more 'public' settings. Of course, many people's use of technologies at home is mediated by other members of the household. On the one hand these others are often considered to be a hindrance or restricting factor to adults' use of ICT – especially in terms of family members. Tellingly, recent market research claimed that 'millions of Britons think their relationship with a computer is becoming more important than time with the family' (*Guardian* 2004: 8). On the other hand, family members often come together to form joint constructions and understandings of ICTs in the home. As Habib and Cornford (2001: 135) observe, 'families and family members engage in rituals and ceremonials, as well as more routine practices, in order to make computer technology less threatening, more habitual and familiar'. Moreover, as we saw in Chapter 5, the households of other people can provide an important source of proxy access to ICTs. The homes of extended family members, relations and friends act as key sites of access to ICT – even for those with access in their own homes (Mossberger *et al.* 2003). The home is also now a crucial site through which the IT industry market the information society vision, with the technologist's dream of the 'smart home' a staple of the IT industry:

> Homes of the future have long been trotted out by tech companies keen to show off their wares, and bridge the gulf between brochure and reality. And the houses – while often leapt upon by TV news hungry for a funny geeky item – have always seemed destined to remain floating in the future. They are normally filled with what the trade calls 'vapourware' – technology long promised but rarely delivered to shops.
>
> (McIntosh 2004: 28)

Above all, the home is considered to be a particularly apposite site where technology and education come together. Indeed, as 'knowledge machines' *par excellence* the introduction of computers into the home has long been heralded as leading to a renaissance of domestic learning. Echoing Simon Frith's (1983) portrayal of early broadcasting technologies leading to a 'rediscovery of the home' as a site for domestic leisure activities, commentators are now eagerly talking about computers (re)establishing the home as 'the place where people do most of their learning' (Tiffin and Rajasingham 1995: 52). Yet, on the whole, these assumptions remain largely ignored by those researchers who have examined the use of ICTs in domestic settings. Existing studies focusing on technology-based learning in the home have primarily concentrated on children and young people rather than adults (e.g. Moran-Ellis and Cooper 2000; Valentine and Holloway 1999; Livingstone *et al.* 1999; Facer *et al.* 2003; Vryzas and Tsitouridou 2002).

Such studies have presented adults, if at all, as parents mainly in terms of financing, regulating and shaping children's educational use of ICTs. They have not asked questions about what learning the adults themselves may actually be using technology for (see also van Rompaey *et al.* 2002; Struys *et al.* 2001; Smet *et al.* 2002).

Although previous literature can tell us little about adults' use of computers for *learning* in the home, a sizeable body of work has grown up around the 'domestication' of ICTs into the everyday lives of adults. A number of media and communications researchers over the last two decades have explored the ways in which ICTs are appropriated and incorporated into the 'domestic sphere', a phrase used to designate what is commonly referred to as the home but also emphasising the extended social, emotional, cultural and political characteristics of the home environment, rather than its purely physical and functional features (Habib and Cornford 2001). Indeed, as Mallet (2004) reminds us, 'home' can refer to a place, a space, a set of practices and even feelings. Technologies are appropriated into the domestic sphere through ongoing processes of gaining possession and negotiating 'ownership', 'objectification' within the spatial and aesthetic environment of the home and 'incorporation' into the routines of daily life (Silverstone *et al.* 1992; Silverstone and Hirsch 1992). Habib and Cornford among others, describes the various dimensions of domestication thus:

- *Commodification*: where the potential consumer imagines, 'constructs' the product;
- *Appropriation*: the consumer buys and accepts the artefact into his or her domestic environment;
- *Objectification*: the artefact finds a physical place within the domestic environment of the consumer;
- *Incorporation*: use of the artefact is fitted into the consumer's domestic time;
- *Conversion*: the consumer signals his or her consumption of the artefact to others (e.g. displays of competence or of ownership).

(Habib and Cornford 2001: 3)

Within this framework new technologies are constantly being interwoven with domestic life (Silverstone 1993). The domestication approach is a useful one as it allows researchers to examine 'how objects move from anonymous and alien commodities to become powerfully integrated into the lives of their users' (Lally 2002: 1) as well as asking questions of how people 'make sense of, give meaning to, and accomplish functions through technical objects' (Caron and Caronia 2001: 39). In this sense it provides an ideal framework for us to move beyond our survey data and gain a deeper understanding of adults' use of computers and, in particular, the influence of family and household dynamics on these patterns of (non)use of computers for education and learning. In particular, this chapter now goes on to examine three questions:

- What role(s) do education and learning play in adults' gaining possession and ownership of ICTs?
- What role(s) do education and learning play in adults' objectification of ICTs within the spatial and aesthetic environment of the home?
- What role(s) do education and learning play in adults' incorporation of ICTs into the routines of daily life?

What role(s) do education and learning play in adults' gaining possession and ownership of ICTs?

The influence of education on adults' commodification and appropriation of ICTs such as computers and the internet was a key part of our interview and case-study data. We know from previous studies that the desire for ownership of ICTs is usually justified via a 'repertoire of official reasons' which are invoked to introduce the technology into the household (Caron and Caronia 2001). In our study, as with others, notions of education and learning were prominent official reasons in the minds of many of our interviewees. However, while there were occasional examples of adults deciding to acquire a computer to meet their own formal educational needs, more often than not this was expressed in the form of 'needing' a computer to assist the education of others in the household – most notably children:

> Pressure from kids and school work. They needed one as well … it was just a general feeling, you know, the kids were after one, um, and you could see the way things were going. We needed one.
>
> (female, 55 years)

> Well I needed it at one stage. I was bothered about jobs for the kids. I knew they'd need the technology and we had to provide that.
>
> (female, 54 years)

Most interviewees citing the education of children as a prominent reason underlying their appropriation of a computer expressed an underlying sense of considering what was 'best for the children' (Haddon 1988). Some of our interviews with younger parents also reflected the 'intense social pressure on parents to buy a computer' noted by other researchers (e.g. Lally 2002; Calvert et al. 2005). As argued by this worker in a furniture factory who had a three-year-old daughter but did not have a computer at the time of the interview, 'it will be a necessity, eventually. She's got to be just a bit older yet … Mainly it's finance … but eventually, we're going to have to, for madam … We just haven't got the money' (male, 43 years). Also recurring throughout the interviews were official repertoires of investing in children's education and, more subtly, gaining a degree of control over children's education. In this way, the computer was constructed as 'a representative of [an educational] world beyond the family' (Habib and Cornford 2002: 132) which could nevertheless be enrolled into the domestic sphere through the appropriation of a home computer. Providing access to a computer in the home was therefore seen by some parents (and also grandparents) as a contemporary form of cultural capital – a notion raised by respondents of different ages and socio-economic backgrounds, as in the case of this mother of a two-year-old girl living in social housing:

> [My boyfriend's] mum bought it for us just when she [the child] was born, so, the Christmas before she was born. So, it was mainly for when she gets bigger and we can then sort of teach her how to use it. I'm sure by the time she's old enough to go out to work that's all there's going to be.
>
> (female, 23 years)

Yet this privileging of new technology as a valuable endowment of educational capital was not consistent across all our interviews, and varied according to adults' own experiences and perceptions of ICT and education. This is illustrated by a young mother of two children from Ely who confessed to 'hating' using computers at school. Also most people she knew who owned a computer used them predominantly for entertainment purposes ('either DVDs or games'). From this perspective the computer was an almost useless addition to the household:

> We had a computer … my uncle bought me a computer but I gave it away and people say 'well you should have kept it for the boys' but they got a PlayStation upstairs so they don't need one.
>
> *Interviewer: Did you ever get it out of the box?*
>
> No I just looked at it and then gave it away, my uncle thinks it is up in the attic but it's not.
>
> *Interviewer: Why did he give it to you? Was it one that he was throwing away?*
>
> No, I must have been drunk and he works for a computer place and he said 'do you want me to pick you one up for you and the boys?'. And I was like 'yeah alright' and then he brings this big bloody box in and on the back of the box it showed you what leads go into where and I thought that 'it just has to go'. The booklet was about 'that' thick and I just thought 'it has to go'.
>
> *Interviewer: Was your uncle really keen on you having one?*
>
> Yeah, he always says 'how you doing on your computer?' and I say 'I'm just reading the book'. God knows what I'm going to say in a year's time.

Interviewer: Has he only just got it for you?

Yes a couple of months back. I will just say the kids dropped it on the floor or something, just blame it on them.

(female, 21–40 years age group)

Whilst children figured prominently in many adults' repertoires of initially stated 'official' reasons regarding the future purchase of computers it was sometimes apparent that more persuasive 'unofficial reasons' underpinned people's actual adoption of computers. For example, as this now retired man explained, whilst initially planning to buy a computer to 'learn with his son', a more pressing influence was the perceived need to keep abreast of skills he was expected to develop for the workplace:

My son and I bought the first one together, oh about 14 or 15 years ago and, of course, I worked for [the council] so we started having computers in the office. So I used to do a little bit of work at home to improve my computer skills.

Interviewer: Were you expected to more or less develop them in your own time?

Well, yes. They gradually brought the computers into the office and then they suddenly decided we didn't need so many typing crew and they allowed us to type our own letters. What I used to do, I used to do quite a lot of work at home because it was so busy, so I often answered umpteen letters in the evening, put them on disk and then took them into the typists to type them out for me.

(male, 72 years)

Also, as in the case of this university lecturer, although education and family members were often cited as 'official' symbolic reasons for adoption, more prosaic (and less memorable) reasons sometimes underpinned the first purchase of a household computer:

Interviewer: In terms of you getting a computer for the first time – why was that?

I expect it was because I had access to a PC at work. Julian – my husband then – wasn't using a computer at work, so I think we bought it, yes the very first thing we bought was – no, I know why I bought it, because I was editing a parish magazine!

Interviewer: Oh right.

So I bought it to edit the parish magazine … but also knowing that the children would start to use it. Then we've upgraded through the years.

(female, 50 years)

Apparent in many interviews were vaguer notions of what Marshall (1997: 71) refers to as 'technophobia of the projected future' and a need to 'keep up' with a changing society. As a steel-worker explained, 'You've got to keep up with the technologies … You've got to be able to work with them' (male, 64 years). Yet often these wider expectations of 'acquiring a handle on the future' through acquiring a computer (Lally 2002) were vague and tinged with ambivalence:

You *need* to have a computer to keep up with everything else and to keep up with technology and with what everyone else is doing … But I don't think computers will push out books and learning that way.

(female, 46 years)

What role(s) do education and learning play in adults' objectification of ICTs within the spatial and aesthetic environment of the home?

From these initial (often educationally-driven) intentions, it has been argued that households and families then go on to construct their technologies in different ways, 'creating private meanings (redefining public ones) in their positioning, patterns of use and display' (Morley and Silverstone 1990: 35). Thus, whilst the public discursive positioning of the computer in many of our interviewees' households was centred around education of sorts, how people privately objectified computers in their homes often had subtle constraining effects on their educational (non)uses. For example, how computers were physically located within the home had significant bearing on their utilisation, collaborating with some household objects and displacing others. We saw in our interviews and case-study visits how computers found their place, space and moment amongst a host of other educational technologies in the household, such as books, televisions and magazines.

For example, in all households there is an ongoing social (re)organisation of the home, where objects are placed and replaced and their functions and understanding of their functions subtly altered, sometimes having a profound effect on the educational uses of objects such as computers. As Jordon (2003: 155) observes, 'space within the home is a constructed element … that – consciously or not – provides family members with a sense of that which is shared and that which is personal'. This was highlighted in the case of the middle aged woman highlighted at the end of Chapter 5. She had left school with no qualifications ('I've got nothing academic') but after buying a computer when her daughter was young found herself learning and retaining ICT skills and other information quite easily. However, as she then described later in the interview, this new site of learning had been used less and less as her daughter had grown older and a spatial repositioning of the computer had taken place:

> I use [the computer] very rarely now. I suppose because it's – it's an excuse, I guess – it's in [my daughter's] bedroom. If it was in the living area, 'cause I've got a small house and nowhere to put it – if it was in the living area I think I would go on and off it quite a bit. Because it's in her bedroom, I tend not to use it so much. And when she first had it, she wouldn't use it all the time. She'd go on it occasionally and I would pop on it now and then. When she comes home from school she always seems to be up there or back and forth. And it's in *her* room, so I tend not to use it so much.
>
> (female, 46 years)

Households would adopt a range of strategies to manage 'access' and 'ownership' of the computer – primarily by managing the spatial confines and location of the computer. Some families would have computers, internet points or digital television sets in communal spaces such as living rooms or 'communal offices' where families could 'all share it' or, more accurately, parents could 'monitor what is going on' (female, 37 years). Some households would have computers in bedrooms, conservatories or cupboards according to perceived function and roles assigned to it (see Figures 6.1 and 6.2). As this woman with her computer located in the kitchen explained:

> Yes, I've got the internet in the kitchen.
>
> *Interviewer: Oh, that's a novel place to have it!*
>
> Yeah! Well it's just, my daughter's got a computer in her bedroom round there; I don't want a computer in *my* sitting room. That's my bedroom – I don't want a computer in *my* bedroom. So it's just tucked in the corner on the breakfast bar, which is quite a good place to have it actually.
>
> *Interviewer: Do you find yourself using it more there, 'cause you find yourself sitting down at breakfast and there it is?*
>
> No, no, no. I'm not that keen on it, but, yeah … Definitely on the internet and phone calls at the moment. I kind of, I know the places, more or less, I need to contact … we get magazines, you know, trade magazines. So you kind of know, you see loads of companies and websites and stuff to phone up for. So that's really how I'm sort of finding out about what I should be doing next year.
>
> (female, 31 years)

Whilst an office or kitchen location was a manifestation of the 'work' or 'informative' roles of the computer, some households would symbolically position

Figure 6.1 The objectification of home computers in two children's bedrooms. The girl's computer on the left was used predominantly for schoolwork. The boy's computer on the right was used for entertainment and schoolwork.

Figure 6.2 The objectification of home computers in two 'adult' spaces. The computer on the left was part of an extensive home office room built up by a male respondent over the past three years. The living room computer on the right was used less often by a middle aged female respondent.

computers to suggest educational uses, such as in 'study' areas or beside bookcases in living rooms, publicly displaying the link between the computer and the stored 'knowledge' of the books. However, interviewees would sometimes privately acknowledge their limited use of these 'educational' resources or express a preference for using the books to learn as opposed to the computer. The disparity between the public positioning of the computer and their actual use in private is highlighted by Mr Palmer who, at the time of the first interview, had acquired a computer and was dedicating a room solely towards its shared use with his granddaughter and their joint learning of ICT skills:

> We're doing some decorating upstairs, we're doing the back bedroom out for my granddaughter and she's going to have her desk upstairs and I'm going to have a desk in there as well then I will get on it then, and play with it a bit more and get on the internet and sort of really sort of push it as far as I can possibly go.
>
> (male, 55 years)

However, upon the case-study visit eight months later it was apparent that no progress had been made with this positioning of the computer in a 'special' space to be used and learnt on. As Mr Palmer ruefully noted, 'well it's *more* or less done at the moment. I mean, I've had a [computer] desk and a chair from Ikea under the bed for nearly a bloody year, and I said to my missus that'll be an antique by the time it's finished'. In the meantime, he continued to sporadically use his laptop computer which was often 'lent out' to his son-in-law who took it away on business trips – thus obviously constraining his father-in-law's opportunity to learn how to use it.

Other respondents had successfully created spaces for their ICT use – although sometimes sacrificing comfort for privacy. As this man described with regard to his loft-space:

> *Interviewer: So is this your private space?*

> Well, it is to a certain extent but I leave the door open most of the time. If they want to shut me out, they just shut the door and watch the telly next door, so it works both ways. It does get a bit nippy in here in the winter and very warm in the summer, so I made these home-made blinds for the summer. In the winter I sit in here with a jumper on and fingerless gloves. It is sort of my private place, it's not entirely private because it's not very insulated, it still comes through the roof but it's secure which is the important thing.
>
> (male, 48 years)

This objectification also extended to the ways in which people organised and set up their file and memory organisation within the computer. Within the desktop environment of the Windows operating system, for example, different family members' uses of shared computers was more easily delineated and 'managed'. During our first case-study interviews which involved an observation of respondents using their computers, it was notable how distinct folders such as 'work' or 'family photos' were often established within the Windows desktop environment (and it is important to note that this was before the widespread adoption at home of Windows XP which creates some of these folders automatically). Rarely, if ever, were dedicated folders set up by adults for learning or educational activities – if they were then it was for children's schoolwork or for informal 'learning' projects such as researching the family tree. Of course, computers are also 'pre-domesticated' before entering the home, with manufacturers and retailers keen to shape the computer via marketing discourses and the physical embodiment of the machine at the 'design/domestication interface' (Silverstone and Haddon 1996). As Silverstone (1993: 285) argues, 'the potential of any technology is not just inscribed within the design of the machine but claimed in the rhetoric of its marketing. It also has to be socially negotiated and learnt'. This was evident in our case-study households by the free 'educational' software that is often packaged with the machine and the commercial 'educational' packages made by manufacturers and vendors. In many of our case-study families such educational software and applications tended to be initially used but then quickly discarded:

> We did try Encarta disc, I enjoyed looking at that.
>
> *Interviewer: Did you think it was as good as the books you read or was it just a novelty?*
>
> I would rather have the book as I love them so would rather sit down and read a book – it's easier to get to it.
>
> (female, 31 years)

What role(s) do education and learning play in adults' incorporation of ICTs into the routines of daily life?

From this brief consideration we can already begin to gain a sense of how the 'home computer' is shaped by the 'already existing practices within the family' (Facer *et al.* 2003: 46), and how this shaping often precludes or limits actual educational use by adults. To understand this shaping further we can also look at the 'moral economies' of households – i.e. the shared/contested views of the appropriate values and practices within the home which are heavily based on the household's class, cultural, economic and symbolic capital (Silverstone 1993). One of the explicit manifestations of such moral economies is the allocation of time to computer use within the household. As Silverstone (1993: 287) observes, 'the order of domestic time is at the same time the product of detailed struggle and negotiation – a domestic politics – fragile, ephemeral, gendered and often the source of conflict'. In this way, households often construct or negotiate explicit 'hierarchies of use' – either institutionalised in explicit house rules and conventions or a series of negotiations. In some of our case-study households with school-aged children the educational patterns and structures of the household were explicitly geared around the children. In some cases, what was privileged as a 'proper' use of the computer within the temporal rhythms of the home was fitted around these pre-existing patterns – a trend, as with Jordon's (1991) study, more prevalent in the households of higher socio-economic groups. If we take the example of the 'democratic' methods employed by this upper-middle class household of Mrs Turner and her two secondary school-aged children – all with competing demands for one shared computer:

> We sit and discuss at tea-time and we look at needs. And if somebody's got an assignment that has to be done that needs to use the web for research, then they have to have that computer. And if I've got a lot of phone calls to make they can't use the web.

> Interviewer: So does this idea of a meeting round the table work?

> It does, yeah. There are times when we have to be fairly strict and say, 'right, you can have it till eight o'clock and then you have to ...'. It has to be like that, I mean, Tim's in the sixth form [Grade 13], Adele's doing GCSEs [Grade 11] and I've got work to do. And we all need to research.

> Interviewer: You discuss this together?

> We three of us work it through. Obviously, when they were younger, it was me. But [now] we actually prioritise their needs, because, you know, if I want to go on the computer, if somebody wants to go on the computer to do something, I want to just check up something – well, hang on, why don't you prioritise that against Tim needing it? But then, the other thing is that I need it for an assignment that's due in tomorrow, but, hang on, you've had that assignment for three weeks. So we get into those debates as well – trying to teach them about how you can't leave it till the last minute and then expect to be top priority.

> (female, 50 years)

As can be seen here, the prioritising of possible uses of the household's resources involved deliberate and explicit negotiation – in this case priorities ran in order from the eldest son's school work, daughter's school coursework and, then, Mrs Turner's work-related use of the computer. However, using the phone for 'essential' household related business took precedence over any use of the internet. Therefore, the mother's use of the computer was a sporadic and fluctuating activity 'as and when I find the time and find it free'. As was demonstrated in the survey data, given the limited time in households with 'competing' users, if adults are to use computers it is mainly for essential 'non-learning' activities such as word-processing and retrieving information from the internet. There were, however, occasional examples in our interviews of high users of computers for formal learning. For example, for the minority of adults who were involved in some form of formal educational activity, the computer played an important role in supporting learning at home. As this woman who had just completed a postgraduate degree describes, her learning was sustained away from the university campus by her relatively 'high-spec' computer:

> So I signed up for the masters, which was a one-year course, and thoroughly enjoyed it. And technology was quite important, because we used to use electronic journals quite a lot and being able to access them from home was – you know, you can have a cup of tea when you're at home, you can't when you're in the library or in the class. So, I found the access to the technology very useful, but I know that some of the students didn't get on with it. And some of the part-time students still can't access electronic journals from home to use later. So I think it's you either can or you can't.
>
> (female, 38 years)

However, across the whole of our sample, adults' educative uses of computers at home mostly involved informal, rather than formal, learning. As in the household survey, we found more evidence of our interviewees engaging in informal learning through computers – some of which could be classified as 'intentional', often sustained and part of a large-scale learning 'project' which people were engaging in. Whilst for some authors such intentional informal learning is seen as being predominantly workplace focused and workplace based (e.g. Lonham 2000; Garsten and Wulff 2003), one of the interesting patterns to emerge from our study was the role of the computer in allowing work-based informal learning to permeate beyond the workplace and into the home setting. As this nurse explains:

> My matron is of the mind, I mean, she likes to know everything, because then you can offer the right kind of care. And this [one patient] has this debilitating condition and he's going to deteriorate regularly and we don't know how to treat him. We don't know what to expect. So we tried all sorts. In fact, I was the only one – on my PC – who found out anything about it.
>
> *Interviewer: On your computer at home?*
>
> Yeah, my friend Jeeves. … It's like this sarcoidosis that I looked at. The front page of the first place I went said, 'when I first got diagnosed with sarcoidosis nobody seemed to know anything about it, so I started this website'. There

you go. Bingo. There were people telling their stories; there was a page where you could see whether there was a help-group near you; there was doctors' descriptions which is what I was looking for, I wanted to know what the hell is it. I mean, I'd looked it up in the medical dictionary and all it said was 'flesh-like'. Right! Ok. So I found this website and it was like I'd started reading the doctors' notes.

(female, 39 years old)

This woman told us how previously she had used to do the same form of work-related home learning with encyclopaedias – only with less success. Thus it is important to note that the computer was not facilitating 'new' forms of work-related learning *per se* – merely reinforcing existing rhythms of learning within the household. As noted in Chapter 5, we also found evidence of similar activities linked to the household, leisure and general interest domains. However, it should also be noted that nearly all these ICT-based episodes of intentional informal learning were seen by respondents as being part of an already established and sustained project that also involved learning via books, courses and social contacts. The multifaceted nature of this learning is illustrated in this extended excerpt from Mrs Sodje, one of our case-study respondents.

Because we are going to Peru I was thinking I would learn Spanish. So I actually bought [a language course] – not one to use on the computer – just one on cassettes. But you can also use the internet … It's BBC … there are free online activities at 'bbc.co.uk'. So I thought I would try that. I did see another one but another reason I bought this is … it is much more colourful. Just the book itself has got pictures in it.

Interviewer: How's it going?

Well, I've only had it two days … I thought I'd do Spanish on the computer but as you can see…

Interviewer: You've got the four cassettes.

Sometimes it's better to do it *properly*. I actually started to learn Icelandic to go there.

Interviewer: Oh my lord, that is a tough language …

I did Latin at school. It was a case of it being useful because I could pronounce the names of places we were going to and I could understand a little bit of things that were said. I'm always a bit wary when you go to foreign places that the couriers are talking and you don't know what is going on.

Interviewer: How long did it take you to even manage to pick up bits of Icelandic?

Um, I can understand why it would be hard. I got a bit stuck when I came to the bit that you have to decline the adjectives but I started in January and we went in June.

Interviewer: So, you've learnt a few languages – the Spanish, Icelandic …

I didn't pursue the Icelandic too much because I knew the only place I could use it was Iceland! Whereas Spanish you can use it in a lot of situations and we may go and spend the winters in Spain as we got older so it might be quite useful.

Interviewer: But you have done a lot of learning over the last year, teaching yourself bits and bobs – but the only thing you went to college for was for the computer course. Would you go to the college to do Spanish?

Well, I thought about that, I was in town on Monday and I seem to remember someone telling me they taught Spanish at the civic centre and I went in to see if there were any brochures and the first thing when I go in 'this building is closed today' and you know? This is kind of typical.

Interviewer: What's interesting is that you could learn Spanish on the internet, there is no need to get books and tapes …

I think it's much easier to learn visually with a book. Some find it hard looking at screens … It's just an irritation; it's not so smooth. I think that's the thing [although] I do use computers quite a lot.

(female, 55 years)

This example highlights a number of important issues. Mrs Sodje was certainly a lifelong learner and, as she intimated, had learnt different languages since school. She was also an extensive user of computers yet her preference for learning a language via non-technological means was pragmatic. She found learning from a computer to be a less convenient and (in her perception) less effective way of learning – not 'as smooth' as cassettes and textbooks. At best ICT was being used as a back-up to learning in the form of the associated website provided by the BBC. Tellingly, with regard to the more formal learning opportunities to learn Spanish, Mrs Sodje's local adult education centre had been less than welcoming although she had considered the option.

As we noted in Chapter 5, also evident from our interviews, but more difficult to quantify accurately, was the range of 'unintentional' informal learning which took place in the home via technology – learning which was smaller in scale and incremental in its effect. Thus, it was clear that for some of our interviewees,

ICTs such as the internet, computers and digital television were contributing in this way to such passive education.

> *Interviewer: Are those things that you know off the top of your head or is it the case of going out and keeping your eyes peeled?*

> A lot of, when I'm asked for an opinion, it is stuff that I know [but] I use Google and this is how I do it ... once you start learning, you know, you keep learning.
>
> (female, 33 years)

> I enjoy reading [webpages] just for the knowledge – I look up history or geography, that sort of thing.

> *Interviewer: So if there was something you wanted to know about and you couldn't find it...?*

> Yes, looking for knowledge and answers to questions. Looking for companies, I'm not interested in that side of it, the shopping thing.
>
> (male, 64 years)

In all of these cases there was little evidence that computers had created a new-found desire for learning – rather that these instances of computer-based informal learning were building upon existing learning behaviours practised in the home. Of course there is a fine line between 'learning' and acquiring random snippets of information, and there were far more instances in our interviews of adults using computers to seek specific information at specific times (e.g. holiday information or sports results) rather than using computers for more sustained quests for knowledge. Indeed, reflecting the patterns in our survey data, the most commonplace instance of learning that the majority of interviewees could report taking place on a computer at home was the gradual learning of computer skills on an informal basis, identified by some of our respondents as 'tinkering around'. More often than not this self-education was expressed in the interviews in mundane and haphazard terms such as 'getting by' and 'muddling along' with computer skills. These issues are explored in more detail in Chapter 9.

Discussion

What the 'home computer' is and what it is used for differs from individual to individual and between households. Although computers may appear similar in their basic forms as physical artefacts, behind the closed front doors of the home they are used:

in ways that are common and unique … their physical position in households, their status as the focus of daily ritual, their incorporation into private and domestic lives will be as varied as the individuals and families who attend, and socially significant (or not) in their patterning and their persistence.

(Morley and Silverstone 1990: 33)

With regard to our concern over what roles home computers play in stimulating and affording adults' education and learning, the simple answer is that the computer, as with all other technological artefacts, tended to fit in with what people already do (and do not do). In terms of education and learning, as with other activities, computers become 'extensions of ourselves, reflections and echoes of who we are, were, and will become' (Romanyshyn 1989: 193). We have gained a sense from our data of how the computer is brought into the household and 'fits in' with pre-existing structures and patterns of education and learning activity – sometimes disrupting and altering (for better or worse) and sometimes reinforcing and replicating already established and entrenched patterns. Belying the rhetoric of the 'le@rning society' the impact of the computer as a 'new' impetus for adult learning was not apparent overall from our data. Indeed, as Lally (2002: 8) observes, if the computer is to 'create' new forms and patterns of learning it is faced with a considerable inertia:

A home computer is brought into the domestic context which is already organised around structures and hierarchies of age, gender and other specific roles, with pre-existing patterns of interaction and activity, and which already contains a large number of objects and other technologies. These structures provide the household's existing patterns of activity with a considerable momentum.

(Lally 2002: 8)

This is, of course, not to claim that computers and the internet were not playing a part in some of our respondents' education and learning. Our findings concur with those of Habib and Cornford (2002: 166) who identified 'learning and freedom' as two of the overriding meanings which adults symbolically attach to computers during the commodification process. Yet whilst it is clear that the computer may appear to have a strong symbolic and status-driven association with their education and learning, adults' adoption and use of computers is often, in practice, enrolled into 'broader projects of domestic development, maintenance and reproduction' (Lally 2002: 11) such as writing parish newsletters, learning work skills and emailing relatives. Whilst discourses and rationales of education figured highly in how our respondents justified gaining possession and ownership of computers it was less apparent in the objectification and incorporation of computers into the routines of daily life. If anything, acquisition of a computer could be seen as acting as a 'bridge' to an imaginary or desired educational future both for children and for adults, whilst its day-to-day use was more prosaic and non-educational.

More often than not, we found that computers in the home allowed people to participate and engage in the types of learning which they were already engaged in. Where some of our interviewees were avidly using the internet for informal learning or self-education, these were people who had previously done so with books, magazines, television broadcasts, friends and neighbours. Moreover, in most cases, these adults continued to learn in these 'traditional' ways alongside the computer. As Morley (2003: 443) concludes, 'even the very latest technologies can always be adapted to suit very traditional purposes'. The use of ICTs for informal rather than formal education and learning is therefore perhaps not surprising when we approach the process of assimilation of technology in the light of pre-existing family habits, norms and values. We know that adult participation in any form of formal education is relatively low and entrenched in pre-existing patterns of formal learning (see Chapter 4). As we have discussed in earlier chapters, we also know that informal learning represents the vast majority of learning that takes place across workplace, community and home settings. In this respect one would expect computers to be domesticated into pre-existing patterns of informal learning rather than leading to the expansion of a 'new' formal engagement with education.

Our interview data reflected strongly the 'technical intermediation' of the domestic sphere as constituting structural circumstances which could be seen as preventing some respondents from otherwise making use of computers for educative and general purposes. We saw, for example, how the spatial demands of the home PC precluded their full integration into the architecturally-confined homes of some of our respondents in lower income groups. We also saw that the complexity of family relationships and household structures were crucial for understanding some female interviewees' (non)engagement with computers (Burke 2003) – especially in terms of the familial negotiations and conflict regarding ownership, control and spatial positioning of computers in the home as well as the 'guilt' of spending time on the home computer at the expense of other members of the family.

Indeed, the mediating role of the pre-existing micro-politics and power dynamics of the household and family was especially noticeable in terms of women's (non)use of computers in the home. Of pressing concern from a social science perspective is the likely self-perpetuating nature of such inequalities. There is a danger, as Resnick (2002: 248) reasons, that inequalities associated with the role of others in an adult's use of ICT 'like many other aspects of social life, is not only produced but reproduced'. As we have discussed, the inequalities which tend to occur throughout the domestication of the computer into the individual's everyday life were often contingent on the pre-existing imbalance of the relationships involved, e.g. between children and parents or between male partners and female partners.

Conclusion

This chapter has illustrated the value of in-depth data in examining the subtle influences behind the broad patterns of (non)engagement and (non)use highlighted in the survey. The data presented in this chapter highlight how seemingly mundane actions (buying a computer 'for the children', putting it in a bedroom) can have profound influences on the ways in which technologies are used or not used in the household. In particular, these data illustrate the ways in which ICTs more often than not 'fit in with' pre-existing ways of doing things – including pre-existing patterns of learning. In this way, expecting home-based ICTs to radically alter what adults do, somehow prompting them into new behaviours, is probably to expect too much of technology. The examples in this chapter have shown how home computers help pre-existing learners engage in more learning. There was little or no evidence of computers creating learning where there was none previously. That said, the home is but one context in which adults come across ICTs. The workplace, on the other hand, is a context where engagement with ICTs is qualitatively and quantitatively different. Use is often coerced, with the individual able to exercise far less choice and control over what they do. How then does adults' engagement with ICTs in the workplace interact with learning?

Chapter 7

Bringing ICTs home

The influence of the workplace

> The workplace is an important part of the relationship between computer technology and learning in working-class life. In the workplace, access to computers is structured by management's legal rights to organise job descriptions, choose work station designs, select the forms of technology, define skill requirements, control advancement, and so on. Because these decisions structure the modes of participation in the workplace, they also go a long way toward defining the mode of learning and the material structure of computer learning more broadly.
>
> (Sawchuk 2003: 189)

Introduction

The aim of this chapter is to explore how the experience of using ICTs at work comes to influence the ways adults use them at home. From other research we already know that the impact of the workplace on family and household life varies with the nature and status of a job (Berg *et al.* 2003), but to what extent and in what ways is this also true for use of technology? Certainly Peter Sawchuk, whom we quote above, believes that the potential impact of work in structuring the modes of engagement with ICTs, and the forms of learning associated with them, is huge. But do our data substantiate such an assertion and are the influences of work-based experiences only limited to working class populations, as he would appear to suggest? Interestingly, given the importance of the workplace as a context of adults' computer use, it is noticeable that 'while several studies adopting qualitative approaches have been published on the use of computers in the home, surprisingly little qualitative research has appeared on their use in the workplace' (Lupton and Noble 2002: 9). Our aim in this chapter is to redress that imbalance by drawing extensively on our own in-depth interview data to explore these issues.

As we have argued in earlier chapters, people's state-of-being regarding their technology use is shaped not only by both their present temperament and motivations but crucially by their life-history of technology use. In this way the world of work is one vitally important context in people's technological careers. It is in the workplace that many people have over the last 15 years or so been encountering information technologies for the first time; for many adults, work has been the 'way in' to ICT use. As Butler (1999: 1) argues, 'today's employees, whether chief executives or office juniors, are being forced to adopt technology to ensure that they can work effectively. In the workplace we are all increasingly touched by technology'. The notion of 'forced' or coerced use of technology is important here. Employees do not own the computers they use, so it might be that their use is more easily often forced, shaped and structured by the environment of the workplace and their employment. As a locus of sustained use of computers, work is potentially therefore a key location where relationships with technology are formed, and it was notable from our survey data that the majority of those who confidently use ICTs in any area of their lives also use them in their work.

However, before looking at work experiences in more detail, it is important to remind ourselves that work is only one 'way in' to ICT use amongst many. As we have discussed in previous chapters, amongst our respondents, family members were often mentioned as influential, with partners, siblings and particularly children often being cited as a reason for first becoming involved in computer use. Another 'way in' was through education; those taking some form of additional training frequently reported being required or supported in learning to use the computer during their course. And a further 'way in' was through some existing passion or interest, which drove the individual adult to explore the computer. The membership secretary of the local football club, the editor of a local history magazine, the games player – in all of these activities there are now opportunities for using ICTs and we met numerous individuals whose passion for them had encouraged them to explore the potential of ICTs in supporting their interest.

However, it remains the case that the most common 'way in' to computer use was through work, and for many individuals, work remained a major factor in structuring their ICT use at home. One-third of respondents to our household survey reported having used a computer in the workplace during the last 12 months and over three-quarters of these reported having used one 'very often'. Out of the adult population as a whole, these represent a considerable number and of course their influence on others once they bring those experiences home, may result in an important 'multiplier effect'. For a considerable proportion of our respondents therefore, work based experiences were highly significant.

Bringing the computer home

For an important minority of our respondents, work and home had, in recent years, become intertwined in ways that made them almost indistinguishable; in many cases, ICTs had been central to that transformation. Mr Mackenzie, a 42-year-old initial interviewee, was a case in point. He was a very heavy user of the computer, working through the screen, he estimated, for about 10 hours a day. He was also geographically mobile – his home was in the south west of England but his office, where he spent two days a week, was in Glasgow and the computer was central to that mobility:

For me personally, I like doing things at home … I mean, I work at home now. I mean, upstairs is my office, I mean. I think nothing of doing something at half-past nine at night or whatever. So I've not got a big hang-up about the division of work and home. I see them as linked. I think people do their heads in actually keeping work at work and home at home, because if you've got something to do late, why go and do it at work if you can just come home at the normal time and do it later.

(male, 42 years)

Computers had helped Mr Mackenzie transform his working practices. He was part of what Mallet (2004) terms a strong tradition of home/work blurring – where ICT is just following wider societal patterns.

Others told a similar if less extreme story. Mr Fitzgerald worked in public relations for his local church diocese; in his work he used a wide range of technology – for word processing, for basic accounts, for developing presentations. Much of this work could now be done at home when necessary. He also used an electronic diary, which was integrated with his home computer; this in turn influenced his home life. As he explained:

[it] is really useful, so that [my wife] knows what I am up to … she could have a look and see what I was doing today. Sunday school in the morning, see the Archbishop in the afternoon …

(male, 34 years)

However, despite the symbolic importance of these individuals as modern, flexible, home-based workers, such people actually represent a relatively small minority of the working population. The overwhelming majority of our respondents, even when they were high users of ICT, had not achieved that flexibility.

Nevertheless, for some people, their experience of using the computer at work had significantly influenced their home use. Mr Kerrigan, a ships pilot, was a good example. He was a heavy user of ICT and even though his work could (by definition) not be brought home, the skills he acquired were easily transferred and he was now an avid home user. The transition from work to home was, he found, relatively straightforward:

Yes. I've got my own computer … I could see the way things were going and I wanted to get on the internet and just be up with the rest of the society, if you like.

Interviewer: And well, was it similar sort of skills that you were using?

Yes.

Interviewer: So you already pretty much knew what you were doing?

Yeah. The internet was a new experience, mind. I hacked my way through that, but it was quite easy actually.

(male, 60 years)

Many others described how they now used ICTs for a whole range of more or less routine activities at home. Mr McDonald worked in a DVD company where the production process was entirely automated and it was his job to maintain the production machines. As he explained, 'my job is totally computer-based' and he brought these skills home in a number of different ways; managing his finances, typing letters. He was also now using his computer to support his part time HND course, some of which he was able to access from home via the internet:

I got a laptop now and I use them all the time to keep my budgets, keep my eye on how much money I'm spending. And with college and things, very handy, with projects and on the net as well with college ... they have this open learning thing, where they put things on the net for you to access all the time, like. Revision and projects and things. That's very handy.

(male, 31 years)

Of course, this is not to say that the intrusion of computers into home life was always welcome. Mrs Wilde, a garden designer, was more than competent in her use of ICTs but her husband used them even more and they were evidently the cause of some family friction:

It's like it's always been there so, it's the third person in our marriage. Me, Peter and the laptop ... We are trying to restrict it, not have him bring it home but it's very difficult.

(female, 48 years)

An even more extreme form of tension was reported by one of our respondents whose neighbours were compulsive computer users. Both father and son had brought computers into the home via work and now sat up all night using the internet. As a result, the amount of money they were spending became a real source of family conflict. Matters reached a head one day and they came home from work to find that Lesley, the mother, had cut all the computer leads into one-inch strips with her pinking shears.

But most of the time it was the positive stories of how work experience had been transferred to home that we were told. One particularly positive story came from Mrs Julian, a 'Learning Support' assistant in a large multi-site FE college. In recent years, she had had to re-skill herself from being a librarian to being able to support young people's learning in a much wider range of ways. Computers were now central to her work – maintaining daily contact with staff in five Learning Support centres on different campuses, ordering books, looking up catalogues, searching databases. But she had also had to learn to use the computer to support students with specific learning difficulties such as dyslexia and it was this aspect of her work experience that she brought home. Both her husband and her son were severely dyslexic, but through the computer, Mrs Julian had found ways to support them. Her son, she had supported through A-levels and into university – checking and re-checking his work, helping him acquire and master a range of learning support programmes – mind-mapping programmes and text reading programmes. And she had also supported her husband in developing his writing skills for the first time in his life:

> My husband's dyslexic but it was never identified 'cos he's 54. [Now] on the computer he confronts his dyslexia and he's editor of the rugby memorabilia society magazine. But you know, guess who does all the proof reading!
>
> (female, 57 years)

Through drafting and re-drafting, Mrs Julian was able to help him take this task on, something that would have been unthinkable for him without her support via the computer.

Once they were competent in their ICT use at work, many of our respondents described how they had brought that knowledge and skill into the home. For some, the impact of their work-based ICT experiences on their home lives was direct, with the boundaries of home and work starting to blur; for others, the influence of work was more indirect with individuals using the skills they had learned at work for other purposes. As we have seen, some of those new uses were relatively mundane – using the computer to keep accounts, for shopping, for email. Sometimes though, those uses were life changing as they were for Mrs Julian and her family.

However, through our research we became aware that not everyone who reported using the computer at work did transfer that experience to home in any significant way; their home-based use was non existent or limited and they had not developed their own skills for learning more. Mr Aitken, for example, was a retired manager who claimed to have used computers extensively during his working life:

> It came very quickly into the computer age then. From early developments, all of a sudden, you know, everything happened to be on the computer, and so it's something that you simply, in management, you took in your stride.
>
> (male, 72 years)

However, now recently retired he had not got a computer at home:

> I haven't got one now. I haven't got one in my home. I don't know what I'd use it for at the moment … it's not something that terribly interests me, you know; I've got other interests that take my time.

We heard a similar story from Mr Scott, who for many years had worked for the Post Office and actually oversaw the introduction of computers into the post and telephone service in his region:

> I ended up as a senior manager. I had a lot to do with computers and their introduction because we started introducing computers into the post office in the early 1950s so we were well ahead and I had a lot to do with them, though in my retirement I want nothing to do with them [laughs]. I haven't got a computer here but I have three sons and a daughter and they all have computers, they are fully computerised so far. If I want anything on the internet I go and visit [them].
>
> (male, 84 years)

What was it, we wanted to know, that encouraged some workers to transfer their knowledge and skills from work to home and what was it that stopped this happening in other cases? As we will demonstrate below, our evidence shows that both work-based experiences with ICTs and approaches to work-based training vary considerably and that this variation can influence the ways in which knowledge and skills come to be 'owned' by workers. We would suggest that it is only when they are actually 'owned' in this way that it is possible to transfer them to the different context of the home. And even then, there needs to be a ready context – both technological and social – for their use in the home.

Generation and changing work practices

When we came to examine our respondents' involvement with ICT at work, it became clear that, whatever commentators such as Butler (1999) might suggest, the experience of using ICTs at work it not ubiquitous; in reality it varies immensely. Many respondents had had no work-based involvement with computers whatsoever. Mrs Gethin, an upholsterer, for example, spoke for a great many of our respondents (chefs, postal workers, shop assistants) when she described the computer as irrelevant to her work:

> *Interviewer: Ok, just tell me about computers. Have you ever had anything to do with them [through work]?*
>
> Nothing at all.
>
> *Interviewer: Nothing at all?*
>
> Nothing at all.
>
> (female, 54 years)

Generation was also important: some adults, who were retired, were too old to have experienced ICT during their day-to-day working lives. Mrs Allen was a recently retired primary school teacher. During her time as a teacher, she had had some experience of using computers in her classroom, but despite successive policy initiatives to encourage their use, practice at that stage had been pretty limited – and in her words 'painful':

> Yes, we all had computers – must be going back about ten years ago – no, more than that now, I suppose – in school. You had a computer in school on which the children printed out poetry or stories, that sort of thing … There were programs we did to help children with reading, they had to do a little puzzle before they could go forward to the next bit of the story sort of thing. But that was very painful.
>
> (female, 57 years)

Not surprisingly, neither of these women had taken up using a computer at home.

Others, like Mr Dando, who worked in the steel industry, had seen several generations:

> Computers came into my job about 15 years ago – there were three generations of them – to begin with all we had to do was enter the measurements that we had taken – and you had to enter it on paper as well – and they were difficult to use. We were given no training but had to simply get on with it – one-finger typing. Over the years we had successive generations of computers and each time the learning process was the same; we had to work it out for ourselves really though we did get some help from the graduates sometimes. What changed over the years was that the computers got easier and we had to use the computer for more and more … so that in the end everything was computer-based; everything had to be done on the computer.
>
> (male, 59 years)

Mrs Upward, a doctor's receptionist, also noted how things had changed:

> Oh yes and funnily enough my mum was a doctor's receptionist so I had a rough idea but the job I do now is totally different from when we first went there. Things have changed within surgeries and the health service in general, we use computers now not typewriters – prescriptions are generated from the computers.
>
> (female, 37 years)

Of course the notion of different 'generations' of technological change is in reality over-simplistic; the speed of change even in the same industry has not been uniform. One initial interviewee, Mr Mears, had been employed for most of his working life as a supervisor in a small engineering company where he had had little or no experience of using computers. He then moved to LG, the high-tech multinational electronics company, where it was simply assumed that he would be computer literate:

> [My old firm were] quite behind the times. When I went to LG's, everything was on the PC, so I had a crash course at the institute to pick up Excel and spreadsheets … It was a mad scrabble it was, to be honest with you.
>
> (male, 36 years)

Mr Mears had returned to his old company after six months but only now, several years on, were they starting to introduce computers into their production processes.

It was also apparent that the change process was not a constant. Mr Schwanauer (male, 33 years), a home-based office designer, had had his professional skills transformed by the introduction of computer assisted design (CAD) 10 years ago but since that revolution, there had been relatively little further change. Although he had to buy a new generation of his design package each year, the changes were limited and he described his technological work environment as relatively stable. By contrast, another of our interviewees ran his own company as a games designer. For him, technology was never static; he constantly had to push at the limits of what he could do if he was to keep ahead of the game:

> I also play games for fun at home a lot, at least twenty hours a week and occasionally I also game with friends … My trying to stay on the cutting edge of what computers are doing, is to an extent, research, so getting, we were talking about getting Wi-Fi here and one of the reasons that I don't mind chucking out money on what may turn out to be a waste of time, is because it's an important cutting edge technology and whilst the business can't afford it and it's got no relevance to the business, it's still, it's a technology thing and I want to understand when I talk about it.
>
> (male, 32 years)

During their interviews, our respondents often gave us some insight into the processes surrounding the changed work practices they described. While some changes were indeed the product of advances in ICTs themselves, in many cases, ICTs worked in combination with other sorts of changes as well. Mrs Julian, whom we described earlier, had moved from librarianship to Learning Support and from managing one library to managing five in different college campuses over many miles. Some aspects of her changed work patterns did seem to be largely technologically driven – changes in cataloguing and new programs to support learning difficulties. But others, such as the need to work across five campuses, were managerially led. Nevertheless, such changes would not have been possible without the support of ICTs; managing five Learning Support centres, at a distance, would have been extremely challenging before the development of email. The same was true for Mr Schwanauer, the self-employed office designer.

The development of CAD systems meant that much of his work was now screen-based. However, the technology alone did not explain how he had become a self-employed home-based worker. This was the result of global changes in management practices that encouraged his previous employer to 'out-source' their office design work and encourage employees to become freelance. ICT allowed that out-sourcing to happen and for Mr Schwanauer to work for much of the time from home. As in Mrs Julian's case, the introduction of ICTs facilitated changed work practices that were not in themselves technologically driven.

Different forms of engagement

Whether our respondents had had the *opportunity* to develop their skills and knowledge in relation to ICT through work therefore varied considerably depending on generation and the ways in which their particular employer had introduced ICTs to work. But whether that experience, even when individuals had had it, *actually* transferred to home use seemed to depend not only on the degree of exposure but on other factors as well. For example, some respondents appeared to have only very limited engagement with the technology. The doctor's receptionist we described above, spent a great deal of time at the computer; however, her tasks on it were limited to entering patients' prescriptions and personal details into a database. She had not purchased a home computer. Mr Mears, in his engineering company, had a similar experience:

> We input figures off production, measurements and so on and so forth … memos and the efficiencies and utilisation of plant and everything, I transferred from handwritten onto the PC. It's just a quality system that you feed the data in and it just sort of analyses the basic data, charts and distributions.
>
> (male, 36 years)

Although Mr Mears had purchased a computer for home use several years ago, he now virtually never used it and even though he was keen to obtain more qualifications, and regularly took evening classes, he had never thought of signing up for an 'online' course. A number of our other respondents were equally constrained – the supermarket maintenance engineer, the steelworker on the production line – both were responsible for using the computer to enter data, but like Mr Mears, their opportunities for interactivity were limited; their role in relation to the technology was essentially passive. Interestingly neither of these respondents were regular home users either.

Others we interviewed were aware of ICTs but considered them unimportant in their work. Mrs Hutchinson, for example, a 39-year-old nurse, felt computers were entirely irrelevant to her work; she could never see them having a place in nursing. Although there was more than one computer in her home, Mrs Hutchinson did not use them to any significant degree. Behind her dismissal of the technology, it seemed to us, was a view that the technology was not central to her professionalism. Indeed some respondents objected to what they saw as the use of technology to limit their autonomy and devalue their core skills. Mr Lennie, a social worker, was one such respondent. In his work with clients, what he considered to be the core part of his professionalism, computers had little or no relevance. However, when he was in the office, he used computers a great deal, an experience he found frustrating:

> The information database has been problematic. I've cursed it left, right and centre. I've complained about it and everyone has too. I've accused the originators of paranoia. But there were so many sections of your screen and you don't have permission to go into it. It was mainly a management tool ... so they can spot, can see what we're doing with our cases and what your performance is. Rather than what we wanted was a whole information database where I can update, produce, do assessments, produce care plans and change them at will. This is the thing that was the biggest bugbear for me.
>
> (male, 61 years)

Others took a similar view. Mr Aitken had, until recently, been a senior manager in a motoring organisation. Computers were part of his day-to-day life in that they were central to the organisation and provided direct links to the police. But although Mr Aitken had a computer on his desk, he did not see it as particularly relevant to how he worked as a manager:

> Certainly from my point of view, so that I could monitor what was going on, I needed one on my desk. But I didn't put it to very much use, you know. I would rather people put me in the picture. I wasn't really high technical in those sorts of fields. My job was mostly making – keeping the organisation afloat, I suppose.
>
> (male, 72 years)

Mrs Lally (female, 63 years) provides another example. She was a highly experienced secretary who used her computer on a daily basis. She had found learning to use the computer easy, 'just a fancy typewriter really', but it did not touch what she thought of as her core secretarial skills; indeed she was disparaging of how such skills were no longer valued.

All of these individuals, it seems, were engaged in a struggle with the computer, which they saw as somehow devaluing or limiting their sense of their own professionalism. Mrs Hutchinson, the nurse, made the point most clearly:

> If a computer came into our work that had a network of medical knowledge for nurses – oh, yes, that would be A1. I'd welcome that with open arms ... Yes, I would love that. Because it opens up a wonderful world. But if a computer come in to say you've got someone who's had a stroke, that means that they need this sort of care – I wouldn't like that – can they eat: yes/no ... Well,

you know, sometimes they can. [or] If it was to quantify your work, then no, I don't agree with it. But if it was to open up a world of other complaints, other people out there with the same things, I mean, wouldn't it be lovely ... Brilliant. ... We could share our knowledge.

(female, 39 years)

For Mrs Hutchinson and all of the respondents quoted above, the transfer of their work-based experience to the home was limited. Only Mr Lennie used a computer at home to any degree and that was only acquired recently; significantly, it was not as a direct result of his work experience but because he was encouraged by his son-in-law.

By contrast many others we spoke to had a very different experience of using the computer for work; rather than seeing their work reduced by the introduction of ICTs, they now saw their work skills taking place *through* the computer. For example, Mr Rees was a civil engineer, managing large construction projects on site. He spoke of his use of the computer in precisely this way:

And I've got a laptop with work, and I use that every day ... I take it to site and I can log on to the network and internet and intranet and all sorts of things, but it's just – I use it quite intensely for programming and planning ... I don't do any actual design, I manage the design process – I sort of steer the design, I suppose.

(male, 32 years)

Mr Schwanauer, as we have already indicated, was a self-employed office designer. Half of his work was undertaken at home in the west of England, where he used CAD packages, and the other half was with his clients, often in London. Unlike say the nurse or the social worker, Mr Schwanauer saw the computer as central to his professionalism. However, he also recognised that there were key parts to his work where the computer was irrelevant:

Interviewer: So is it exactly the same, as you'd be using a PC for if you were in the office?

Pretty much. I've got the same software on as I'd use at the office. It's all Microsoft based, all Windows based, so it's easily transferable, compatible ... I can work on a drawing here and send it in ... What I can't do is talk to people really. The phone is fine if you're after something specific but a lot of being a project manager is really face-to-face, you've got to talk to people rather than just yell at them over the phone.

(male, 33 years)

Mr Kerrigan, the ships pilot, gave a similar account:

Yes. We had computerised plan maintenance systems. I worked off a laptop because when I travelled as superintendent you used to do your reports on the hoof sort of thing, you know. But you could communicate from your desks with ships by email, by fax. You could pick up the phone and speak to a ship anywhere in the world if you wanted. As long as you knew which satellite it was working. So yes, in the end it was a huge part of my day – I'd spend the whole day working from a computer at the end of my career.

(male, 60 years)

Many of these people did make the transfer; they used what they learned in work and brought it home, sometimes for routine activities like managing their accounts or shopping, and at other times for more complex activities such as working with family members and (very occasionally) as support for more formal learning. Their actual work-based experiences were immensely variable but what was clear was that they had had the opportunity to develop 'ownership' of the skills they were required to use at work. Work-based experiences, in themselves we found, were unlikely to transfer to the different context of the home if they were not 'owned' in this way. And as we have seen, some work-based tasks were structured in such a way as to facilitate or to prevent real ownership. But what was the role of work-based training in this process; how did training support or inhibit the development of 'ownership'?

Learning and training for ICT at work

From our interviews, it was apparent that those whose main 'way in' to computer use was through work, potentially had at least three different types of learning experience: courses, direct practical experience and networks of support. Each of them contributed to learning in different ways. For example, some of our respondents had had substantial formal training in computing. Mr Mackenzie, the flexible home-based worker whom we described at the beginning of this chapter, learned during his degree:

I had to learn it to do my job; it was part of my apprenticeship. After that, while I was working for the next five years, didn't really use computers in my job at all. Everything was recorded in paperwork – wasn't entered on screen. Then when I went to do my second degree, it was – software was coming out so I did a lot of software. … So from then, computers, computer programming, you know, from the work point of view I've used them and, you know, at home.

(male, 42 years)

As we have already noted, generation is a fundamental factor in influencing the experience of computing and however good Mr Mackenzie's original training it was not relevant at the beginning of his career. Mr Schwanauer, the office designer, also came into his career too early for ICT to be part of his professional training though as he observed, contemporary professional training in office design is now entirely computer-based; as he put it, 'technical drawing is dead'.

In reality, only a small minority of our respondents reported formal training in computing prior to their work. A number of others, however, undertook generic training such as the European Computing Driving Licence. Mrs Dennis who had had a variety of jobs – accounts clerk, working in an estate agent's office – was currently preparing to go back to work after taking time out to have children. She was just completing the ECDL and was thoroughly enjoying it, so much so that she had become an instructor:

> I'm now kind of a teacher – it's the best job I've ever had, I love it. Although it's evenings and it's a pain having to wait all day to go to work I just love it, I'd go in even if they didn't pay me!
>
> (female, 29 years)

Mrs Dennis was finding a use for everything in the course – through teaching it. Others reported that this sort of generic training was problematic in that it was not always relevant to their work-based needs. Mrs Upward had difficulty with the ECDL.

> I started to do the European driving licence but I don't think that was very successful, the only thing I understood and got on with was the database, I probably would have got on better with Word but I don't use it a lot … I don't have a computer here [at home] and as I say I don't actually use Word within my job so really I didn't get the practice on it. I know how to copy and paste but when you are thrown in the deep end and you got this paper in front of you and you haven't done an exam since your O-levels it's a shock – I went to bits and was shaking.
>
> (female, 37 years)

What became clear from our respondents was that what made for effective training was not the length of the training itself (many individuals who had successfully transferred their experience from work to home had had little or no training) but whether that training was relevant to the actual work they were undertaking at work. Only in this way, it seems, were the skills they were taught, consolidated and owned. Mr Benn was a case in point. He began using computers in the steel works, simply entering data on the production line, but then, as part of preparation for redundancy, he was offered training in programming:

> I did a programming course at work – Mallard basic. It was six months they trained us all and it was redundancy time and they wanted us to swap departments so they put us on this course for a year … We were actually writing the programs.

Interviewer: Did you manage to use that stuff that you'd done on the course again?

No, never again. It is just a fading memory.

(male, 56 years)

Despite his substantial training, therefore, Mr Benn's skills withered once he returned to the workplace where they were not relevant. Others on government backed training schemes in our sample reported similar experiences; however good the training, if it was not relevant to their actual work experiences, then their newly acquired skills soon faded both at work and at home.

Others who went on short-term courses specifically tailored to particular jobs by and large spoke of them more positively. Mrs Scott, for example, was a police typist:

For me it's a need to know basis; show me what I need and don't overload my brain. There are lots of different systems and I'm familiar with aspects from each system, it's enough for what I need to do …Well we've got a training pod in the station now, so some of the training is there, some used to be down in HQ. There are specific courses that we do.

(female, 37 years)

Mrs Radcliffe, a supermarket worker, had a similar view:

They took us on a training course and we had to go through the way you had to do it and they showed us how to get back. If you did anything wrong how to do it right. We had three days training before we went and basically did it by ourselves and then learnt by our own mistakes! … but it doesn't always work when you have been trained! You cope and then you learn and it's all right.

(female, 41 years)

What Mrs Radcliffe highlights is the importance of practical experience, whatever the nature of the training. And many of our respondents reported surviving perfectly well with no training whatsoever – only practical experience. Mr Kerrigan, the ships pilot spoke for many when he explained:

I don't think I ever went on a training course, as such. It was just sort of, it grew – on all of us, in a way.

Interviewer: And how do you end up actually learning it? Just sort of on the hoof?

Yeah!

(male, 60 years)

Others did not experience things changing slowly; they were simply thrown in at the deep end. Whether they survived, depended on their own ingenuity and who was around to help them. This was Mrs Dodgson's experience:

Yeah, I suppose my first exposure for computers was putting in information and setting up spreadsheets because we devised competitions for sales agents ... So I hadn't had any tuition at all and they sat me in front of this computer and said you have to do this!

Interviewer: What did you do if you pressed the wrong button?

Say 'help!' ... and I always knew there were particular people who I worked with that I could approach and say can you help me. They'd help me but there's other people you'd say to and they'd say sort it out yourself, you're on there, do it, and no help. I suppose I'm a weakling in that way that I have to ask.

(female, 52 years)

But as Mrs Dodgson makes clear, if relevant practical experience was essential to develop and consolidate computing skills, so too was support. Virtually everyone we spoke to seemed to have someone to turn to at work when learning through practical experience. Mr Kerrigan, the naval pilot, again:

In the early days, there was always someone, whether he be in Purchasing or Accounts or whatever, he was the computer guy. You just sort of, 'John, what do I do about this?'. He comes along.

(male, 60 years)

And even our most highly skilled respondent, the games designer, at times felt the need for support. On these occasions he would telephone Mark, a friend who worked in the computer lab at a local university:

Well at work I write about a hundred or so emails a day to customers and to business colleagues. I use the computer to process the games that are for money. Everyone in the office needs a computer to do things, so I also maintain their computers and that ... [but] I'm not at the top, I might seem compared to your sister ... but as I got friends who are pros, I know there's people who are, you know, higher than me on a food chain.

(male, 32 years)

Steve and his partner Ms Hawkins went on to explain how this informal teaching process worked; it could be characterised as a kind of 'guided support':

Ms H: Steve taught me the basics of HTML and probably various other things as well. They sort of, things like that I've picked up from Steve and I've either run with it and sort of, learnt a bit more myself or not.

Interviewer: And what kind of teaching was it? What did he actually do? Did he sit down at the computer and go over things?

Ms H: Well with the HTML we just sort of spent a couple of evenings, you said, 'this is what you do', show me some, some of the basic guides and wrote down some others and after that I bought myself a book and I think in the end basically we didn't, we didn't go any further than the basics, it was just a question ...

Steve: And then you might ask a few questions and after then a few months, you were learning as much as I did.

During the development of her computer skills, Ms Hawkins had taken some basic training in Access and Excel but more importantly she had both a practical context in which to experiment and develop her skills and she had access to support when she needed it. Through this process she had been able to consolidate her ICT skills and knowledge and to 'own' them; this in turn gave her the confidence to transfer those skills to new contexts including her home life. Many of those who had successfully transferred their work-based ICT skills to home reported a similar pattern. Whether or not they had had formal training, their experience of ICT at work was one which supported rather than undermined their core skills and they inhabited a learning environment that encouraged experimentation and offered support when it was needed.

Discussion

What have we learned about the ways in which ICT use at work provides a 'way in' to home-based use? Certainly work experiences were an important entry for many of our respondents and as we have seen, the influence of ICT-using workers in the home was often multiplied as they came to influence partners and children and were able to offer advice and support to other, more distant relatives and friends. Work it seems is a significant driving force in the domestication of new technologies in contemporary society.

But we have also learned much about the variation in workers' experiences; how exposure to ICTs has varied through time and even today varies dramatically between different industries and employers. We also noted the different ways in which ICTs at work allow and encourage different forms of engagement with technology, some of which employees find empowering and others of which are seen as controlling and undermining of their skills. These differential forms of engagement, we found, have a powerful influence on the degree of ready transfer of knowledge and skills to the home. Finally we looked at work-based learning; this too was immensely variable. Again, what was critical if employees were to be able to transfer what they had learned to the different context of the home, were learning experiences that allowed them to 'own' their new skills. Whether or not formal training is offered, employees need a practical work context that is supportive of experimentation and able to offer guided support when it is needed.

But of course, even when all of these conditions are met, even when individuals have the opportunity to develop and own their new skills and knowledge, there is no guarantee that they will definitely transfer them to the new context of home. For that to happen, there needs to be a ready context – both technological and social – for their use in the home. Some work-based skills are simply not relevant – computer packages to design bridges or to navigate ships or to catalogue libraries. Moreover even when some aspects of those skills are transferable in principle, there still has to be some purpose in bringing them home. Mr Hopkinson, who used a computer regularly at work and had bought one for home, seldom used it; he much preferred to use his Playstation to play games and, as a young single male living at home with his parents, he found little other need in his personal life to use the computer. Mr McDonald was similar; although he used his computer to undertake, what he called, routine, day-to-day activities, it was not particularly important for him. Perhaps this was because he was single and because most of the friends he socialised with were builders who did not use computers at all. To return to the example of Mrs Julian, the librarian, what our case study of her and her family shows is that she did not bring home her technical skills for cataloguing or ordering books online; what she brought home were her skills for supporting dyslexia and she did this because she saw a real purpose for it within her family life.

Before experience of ICTs at work can be harnessed for social purposes such as supporting lifelong learning, there are many conditions to be met. People need the sorts of work-based experiences and learning opportunities that will allow them to develop real ownership of their new skills and knowledge; currently those sorts of experiences are not distributed equally across the workforce. Even more important is the recognition that just because an individual has developed the ICT skills that could allow them to engage in further forms of learning, there is no guarantee that this will happen. Just as with more conventional forms of lifelong learning, what is needed is some reason for undertaking further training in the first place. For the overwhelming majority of our respondents, there was little evidence, even when they had, through work, gained the sorts of skills that would allow them to exploit the new opportunities for online learning, that having those skills in themselves made them any more likely to want to take up new training opportunities.

Chapter 8

Adult learning with ICTs in public and community settings

Introduction

Although the home and workplace are the sites where adults are most likely to come into contact with ICTs, in terms of the establishment of a socially inclusive le@rning society perhaps the most significant setting is the very public provision of ICTs in locations such as libraries, museums, colleges and community centres. As we outlined in Chapter 1, providing access to ICT outside home and work is seen by many policymakers as the only way to reach many of the social groups least likely to be engaging in learning. These are the least likely to be in a job involving meaningful use of ICT (or even be employed at all) as well as being less likely to have a home computer. As there is little chance of this situation being addressed through the marketplace it has therefore fallen to the state to facilitate 'universal' access to ICTs for those outside the relatively narrow, commercial concerns of the private sector. To date this obligation has been fulfilled principally through the provision of 'shared' or 'open' access to ICT in public sites, such as libraries, museums, colleges, schools and in purpose-built sites, designed to complement and extend the longstanding public provision of ICTs by other organisations, such as commercial 'shop-front' facilities and community groups.

Despite countries such as the UK now boasting thousands of public ICT centres there is a lack of sustained empirical evidence relating to their effectiveness in terms of stimulating either ICT use or ICT-based learning. As Devins *et al.* (2002: 942) conclude, there is 'little evidence in the public domain associated with the learning outcomes or impact of such centres in relation to skills development, providing access and in tackling issues associated with digital exclusion'. Although, as we saw in Chapter 1, there are suggestions that such patterns of usage are leading to the inclusion of 'minority' groups of ICT users, there has been little generalisable mapping or detailed examination of usage.

Some key issues, therefore, remain unexamined with regard to the 'effectiveness' of the current provision of public ICT sites in the UK. For example, we have yet to develop an adequate understanding of the patterns of engagement with the public provision which is currently on offer and the real extent of its 'inclusiveness'. More sophisticated questions also remain as to how the placement of centres in different types of settings influence how they are (not) used. For example, what is the contribution of social setting to the take-up of public ICT? What are the unique factors in different social settings and how do the characteristics of different social settings 'configure the user' as well as 'configure the technology' (Woolgar 1991)? In line with our approach in previous chapters there is a particular need to develop detailed understandings of the *non*-users of public ICT sites. How many non-users of public ICT sites have the means to access sites if they wish but simply choose not to do so? How many do not have the opportunity, skills and connections and are prevented from using? If people's technology needs are not being met by public ICT provision then developing a more detailed understanding of what these needs are and how they can be addressed is imperative. This chapter, therefore, offers a detailed look at the current levels of use and non-use of public ICT sites throughout our population of study. In doing so it revisits our survey and interview data and focuses on the following three areas of questioning:

- Who is using (and who is not using) what forms of public ICT provision?
- How and why are adults making use of public ICT provision and with what outcomes?
- Why are more adults not making use of public ICT provision?

Who is using (and not using) what forms of public ICT site?

As we saw in Chapter 5, low levels of engagement with public ICT sites were evident within our household survey sample. Eleven per cent of respondents from the initial survey sample of 1,001 reported making use of computers in a public ICT site during the past twelve months – as opposed to 44 per cent making use of ICT at home and 32 per cent making use of ICT in the workplace. Most of those respondents making use of computers in public sites had done so in local educational institutions such as schools or colleges (5 per cent) and libraries (4 per cent). Only 2 per cent of respondents had made use of computers in community sites and 3 per cent of respondents in either museums or commercial 'pay-per-use' sites such as internet cafés. This is not to say that these modest levels of use were due to a lack of perceived access to public ICT sites. Indeed, 34 per cent of the survey sample reported having *potential* access to some form of public ICT site – with most of these respondents citing libraries (28 per cent of the sample), commercial pay-per-use sites (14 per cent) and local educational institutions (10 per cent) as offering access to ICT if they needed it. In this respect more respondents cited a public ICT site as providing potential access to computers than cited their workplace or place of study. Yet, in terms of supporting *actual* use of ICTs, public sites were substantially less likely to be cited than the home or even the homes of friends and relations. This discrepancy is intriguing.

In order to gain a more detailed picture of who was making use of public ICT sites we can examine the survey data which we collected from a 'booster' sample of 100 known users of public sites, thus augmenting the number of public site users in our survey sample to 208 (see Chapter 3 for a description of this sub-sample). As can be seen in Table 8.1, within this enlarged sample there were noticeable demographic differences in respondents' use of different types of public ICT sites. Perhaps the most pronounced disparities were evident in terms of the age of respondents – with younger and middle-aged adults more likely to use computers and the internet in libraries, local educational institutions and museums. Younger adults were also more likely to have made use of ICT in 'community sites' such a community centres or church halls. The only site of public ICT access not displaying a marked age differential was the 'commercial' category (i.e. internet/cybercafé).

Patterning was also apparent in terms of respondents' socio-economic status. Respondents from the 'service' group, for example, were more likely to have made use of public ICT access in museums/science centres and local educational institutions. Respondents in the 'partly skilled' group were more likely to have made use of ICT in community centres and noticeably less likely to have used ICT facilities in libraries, commercial sites, museums or local educational institutions. Respondents in the 'partly skilled' group were also less likely to have used computers at home or at work. Finally, although gender was not a significant factor in terms of whether a respondent had made use of a public ICT site during the past 12 months, there were slight sex differences in the type of site used. Women were more likely to have made use of computers in educational institutions such as schools or colleges. Men were slightly more likely to have made use of a community ICT site.

It is clear from this brief consideration of the boosted survey data, that the type of public site usage amongst our sample is significantly patterned across different social and economic characteristics. In order to investigate the nature of these inequalities of engagement further, Table 8.2 compares a wider range of social and economic characteristics against the frequency of respondents' use of public ICT provision (i.e. in terms of 'frequent users' who claimed to have made use of public ICT sites on a 'very often' or 'fairly often' basis, 'occasional users' who reported making use of a public ICT site only 'rarely' and respondents who reported themselves to be 'non-users' of public ICT sites).

Whilst differences are evident in these data, they suggest a more complicated patterning of individuals' engagement with public ICT provision than was previously apparent. Some characteristics showed no pattern. For example, use of public ICT sites (whether on a frequent or occasional basis) did not differ according to respondents' sex. In other instances, differences were only apparent in terms of one of the categories of 'frequent' or 'occasional' user. Whether or not an individual was a frequent user of public ICT sites was not found to differ according to socio-economic status, although occasional use was heavily skewed in favour of the service group (16 per cent) as compared to the partly skilled group (5 per cent). Differences were found in relation to frequent use and respondents' ethnic group, with a significantly higher proportion of 'non-white British' respondents making frequent use of a public ICT site than respondents who classified themselves as

Table 8.1 Levels of use of PCs/computers in different locations by age, socio-economic status and gender

	Library	Local educational institution	Commercial 'pay-per-use' site	Community centre/site*	Museum/science centre	Workplace/place of study	Friend or relative's home	Your home	Sample size
Age group									
21–40 years	9	7	3	8	5	48	28	61	383
41–60 years	10	9	4	4	4	46	17	57	351
61 years or more	4	3	3	2	1	5	5	22	367
Socio-economic status									
Service	9	13	2	2	10	53	22	76	88
Skilled non-manual	11	8	5	3	5	48	21	59	334
Skilled manual	9	5	4	5	3	32	19	62	100
Part-skilled	5	5	2	6	2	20	10	31	456
Other	6	7	2	5	2	29	25	42	123
Sex									
Male	7	5	3	6	4	33	17	51	449
Female	8	8	3	4	3	33	17	44	652
Total	7	6	3	5	4	33	17	47	1,101

Notes
* 'Community centre/site' includes community locations not covered by other categories – e.g. community centres, halls and resource centres, village halls, parish halls and church halls.
Data are percentage of respondents in the boosted sample (n = 1,101). Summed data may not add up to 100 per cent due to rounding.

Table 8.2 Usage of computers in public ICT sites by personal characteristics

	Frequent user	Occasional user	Non-user	Sample size
Sex				
Male	9	8	83	449
Female	10	8	82	652
Age group (years)				
21–40	14	9	77	383
41–60	10	11	79	351
61 or more	5	4	91	367
Marital status				
Single/separated/widowed	10	5	84	393
Married/living with long-term partner	10	10	81	686
Health status				
Long-term illness/disability	7	6	88	241
No long-term illness/disability	11	9	81	849
Education				
Continued after 16 years	12	13	75	425
Completed education at or before 16 years of age	8	5	87	676
Socio-economic status				
Service	10	16	74	88
Skilled non-manual	10	11	79	334
Skilled manual	10	9	81	100
Part-skilled	9	5	86	456
Other	11	4	85	123
Ethnic background				
'White-British'	9	8	83	1,002
'Non-white-British'	19	8	73	99
Household composition				
Single adult aged 16–59	19	5	76	150
Small family	12	12	76	388
Large family	9	9	82	182
Large adult household	16	11	74	19
Adult aged 60 and over	6	5	89	203
2 adults, 1 or both aged 60 and over	2	3	95	159
ICT experience*				
Had (had) a job involving the use of ICT	12	15	73	488
Never had a job involving the use of ICT	21	7	72	217
More than five years' experience of using ICT	12	14	73	527
Less than five years' experience of using ICT	24	6	70	178
Used a computer at home during the past 12 months	14	16	70	429
Not used a computer at home during the past 12 months	16	6	77	276

continued…

Table 8.2 continued

	Frequent user	Occasional user	Non-user	Sample size
Location**				
Beaufort	5	4	91	78
Canton	10	7	83	84
Chew Valley	2	6	92	83
Cinderford	8	9	82	85
Cyncoed	4	2	94	83
Ebbw Vale	8	3	90	79
Ely	1	1	98	84
Hartpury	5	14	81	77
Lansdown	4	11	86	86
Nant-y-glo	0	1	99	91
Radstock	1	12	87	84
Tidenham	3	9	87	87
Total	107	86	906	1,101

Notes
* Not including the 396 individuals who had never made use of a computer during their lifetime.
** Not including the booster sample of 100 public site users given its regional imbalance.
Data are percentage of respondents in the boosted sample ($n = 1,101$) except where stated. Summed data may not add up to 100 per cent due to rounding.

'white British'. However, the fact that there were only 99 non-white British respondents in the sample makes this a tentative finding. Other characteristics were found to have patterning across frequent *and* occasional use, such as respondents' educational background, age and household composition. The issue of prior experience of ICT being a mediating factor in people's use of public ICT sites was also highlighted in these data. Respondents who had less experience of using ICT (e.g. those respondents with no workplace experience of using ICT, less than five years' experience of using ICT and who had not used a computer at home during the past 12 months) were more likely to be frequent users of public ICT sites. Conversely, experienced ICT users were more likely to be occasional users of public ICT provision.

Unsurprisingly, differences were also found in whether or not an individual made use of a computer at public ICT sites in relation to geographical location. For example, frequent use of public ICT sites was most prominent in Cinderford, Ebbw Vale and Canton. As we described in Chapter 3, these communities are either in town or city locations, with medium to high levels of economic and educational deprivation. All three communities were also home to a range of public and private ICT provision. On the other hand, frequent use of public ICT sites was least apparent in Nant-y-glo (0 per cent), Ely (1 per cent), Chew Valley (2 per cent) and Radstock (1 per cent). These four communities are less homogenous. Both Ely and Nant-y-glo can be considered 'deprived communities' with some of the highest levels of economic and educational deprivation in our sample. But in terms of public ICT provision Ely was the most well-served community in our sample whilst Nant-y-glo was one of four communities with no immediate provision. Chew Valley also had no immediate provision but was one of the most prosperous wards in our sample in terms of economic and educational activity. Similar patterning occurred with regard to occasional use of public ICT sites. With regard to the proportions of non-users of public sites the same conclusions can be drawn – with some of the 'most' and 'least' deprived communities displaying the highest levels of non-engagement, and 'medium' communities with a variety of public ICT provision displaying relatively higher levels of use. Local provision of ICT centres is a prerequisite for use but, as Ely shows, is not sufficient in itself. Social and economic differences also play a part.

Several conclusions can be drawn from these survey data. First, usage of public ICT sites is low in relation to the use of ICT in the home and workplace, as well as in relation to the general levels of non-use of ICT amongst our sample (even among those with reported access). Second, the low levels of public ICT site usage are patterned along the lines of socio-economic background and age (although not sex). Moreover, within these general patterns of inequality, different types of public ICT site were being used by different social groups. For example, whereas younger adults from lower socio-economic groups were more likely to be making use of ICT provision in community sites, individuals from higher socio-economic groups and with a more middle-aged profile were more likely to be using ICT provision in libraries, museums and local educational institutions.

These findings raise questions relating to the nature of these different types of sites and their effect on the take-up of ICT facilities. It would also seem from the survey data that the frequency, as well as the type, of use of public ICT centres is significantly patterned. Frequent users of public ICT provision could be characterised as younger, from a range of socio-economic backgrounds, with little experience of ICT (and, tentatively, perhaps more likely to be from 'non-white-British' ethnic groups). Occasional users, on the other hand, were more likely to be older, relatively educated, from higher socio-economic backgrounds and with more experience of ICT. These patterns raise questions as to how these different groups are using public ICT centres and with what outcomes. Finally, differences were also evident between the communities themselves – patterning only partly related to the levels of public ICT provision in each area.

In order to further explore these patterns of (non)use of public ICT sites, we now go on to interrogate the data collected from the follow-up in-depth interviews. In particular we can use these interview data to explore the experiences, circumstances and motivations of the three categories of 'frequent', 'occasional' and 'non' users of public ICT sites and, therefore, address the remaining questions of how and why are adults making use of public ICT sites, with what outcomes, as well as why adults may not make use of public ICT sites.

How and why are adults making use of public ICT sites and with what outcomes?

Within the minority of interviewees who *had* made use of a public ICT site a distinct division could be drawn between those who had only limited experience of public ICT sites (often in the form of a discrete and short-lived set of episodes) and those interviewees who were making (or who had made) extensive and sustained use over a period of time.

At the time of carrying out the interviews only seven of our interviewees could be said to be (or to have been) extensive users of public ICT sites. All of these interviewees could be seen in terms of what Cook and Smith (2002) term a 'life cycle' of engagement with ICT centres, involving initial contact and use developing into more regular engagement and often quite significant outcomes. This is illustrated by one of our initial interviewees (Mrs Smith) who had been using her local college-affiliated ICT centre for the past three years and, in doing so, had progressed from being a non ICT-user and non-learner to being a competent (if not always confident) ICT user and part-time tutor. Mrs Smith was a 38-year-old single parent living on a council estate on the periphery of the rural market town where she had lived all her life. She left school at 16 to work as an administrative assistant at a local Ministry of Defence establishment. There she had worked as a typist but had not come into contact with computers; 'I had a family before I had a chance to work on the VDUs'. After 15 years – having had three children and latterly gone through a divorce – Mrs Smith decided to attend women-only returnee sessions at her local college 'mainly to build up my confidence'. Although wanting to take courses in animal care and horticulture she had been advised to enrol on an 'IT skills for business' course in the college's ICT centre due to timetable clashes:

> I was told if I wanted to do animal care or horticulture I would have to wait until the following December. In the meantime, the teacher that was taking you on the Second Steps [course] was worried that I would lose confidence – it took an awful lot for me to go on that course – and she said, 'if you go home we usually find women don't come back'. So I said, alright, if I do the other [courses] that'll keep me going ... she advised me to go on the IT course because computers are used everywhere anyway. And as I could already type I thought, well why not?
>
> (female, 38 years)

From this unintended start she had continued with courses at the ICT centre for the next two and a half years – building up a portfolio of skills, developing a network of friends and eventually gaining part-time employment as administrator and then occasional tutor in the now re-branded learndirect centre:

> After I'd been at the college for a while, we were sat in the internet café having a coffee break and we got quite friendly with the lady, the manageress

down there, and she mentioned that she could do with someone to help out with the refreshments so she could get on with helping the people with the computers. And my friends all volunteered me. … So I started helping at the internet café and through that the lady who was the manageress at the IT suite said, 'would you be interested in work experience once a week next year on your course in my department?'. So I was doing the internet café once a week and the IT suite once a week – mainly admin. And through that, I then went down when both amalgamated [into the learndirect centre] so I went and helped in there. And through that I've gradually gone and become a member of the teaching staff.

Mrs Smith was then working as holiday cover for the ICT tutors in the learndirect centre as well as continuing to take her own qualifications; 'I've come from not knowing a computer to actually helping others. I get a lot of satisfaction from it'. This official position of expertise belied her non-use of computers outside the ICT centre; she did not have either a telephone or computer at home and had to walk two miles into the town in order to use the ICT centre's computers for her own studies. As Mrs Smith said, this was an increasing difficulty, although at the time of the interview the manager of the learndirect centre was arranging to give her an refurbished computer from the centre for her home use.

Cook and Smith's (2002) notion of the 'life-cycle' of community ICT centres refers both to their users and those who work in the centres – acknowledging that transitions such as Mrs. Smith's from learner to tutor are not uncommon. As Mrs Smith's case demonstrates, public ICT sites are at their most effective when they act as a means through which individuals can initiate and augment their ICT 'careers', as well as educational and vocational careers. Once ensconced in the context of the ICT centre Mrs Smith's progress was evident, although she had found making the initial commitment to go there difficult:

> I'd heard about it and I saw it advertised, and I put it off and put it off – for about a year I was thinking about it, before I eventually thought, 'go up that college, put your name down and see how you get on with it – no one can force you to do anything'. So I did that and just went along with the flow. I was nervous. I wanted to back out lots of times, but I just stuck with it. And I'm glad I did now.

In this respect Mrs Smith, as with our other frequent users of public ICT sites, could also be characterised as an archetypal 'returner' – i.e. an individual rectifying previous cessation of their career and/or education (see Edwards 1993; Blaxter and Tight 1995; Selwyn and Gorard 2002). As Hughes (2002: 415) observes, 'in terms of return, the most basic understanding is that of a return to the unfinished business of education'. The role of public ICT centres in supporting such returners was illustrated by another of our initial interviewees (Sarah Jones, a young mother of 22 years) who explained that computer courses were the highest-profile and most potentially useful courses on offer once she had taken the decision to return to education and 'get something behind me':

> I moved out of home, so, and then I had my little girl so I decided I'd take the opportunity to sort of go back to college really. So I went and do some computer courses ... I wasn't then working full-time – it was only factory work I was doing – so then I decided I'd give up work full time, I was only going to go back part-time anyway, so I thought I'd take the opportunity in the daytime to put her in the crèche and, you know, actually learn something.

Since taking an introductory course at the local learndirect centre, Sarah had then gone on to take three progressively more advanced ICT courses and was now planning to take a business skills course. She felt successful, having gained confidence and work skills as well as progressing from a limited base of ICT skills to familiarity with a range of computer applications, 'I mean, I didn't have a clue how to do databases before ... and it does give you more confidence'. In particular, Sarah pointed towards the social context of the learning centre as being 'apart' from her previous experience of schools and colleges:

> You can just book your own times and days in, so you then you just go at your own pace, which is quite good, you don't have to – I mean, you're not pressured that you've got to get something done by a certain time.
>
> [...]
>
> I mean, there's still the same tutors down [the learndirect centre], I think it's just a nicer environment altogether really, just the way it's set out and everything. And because, when I was at the college before, sometimes you'd have some of the kids in there from the college ... the teenagers or whatever ... but now you get a lot of older people down there as well. So, I don't know, I just find it a lot ... you don't need to walk round the college ... Sometimes it can be very intimidating when you're walking around and you've got a big group of teenagers sort of by the doors – even though, like, you're a lot older, it is very intimidating sometimes so. I just find it pleasanter really; it's more pleasant.

For the other interviewees who were classed as frequent users, use of public sites was focused more around the autonomous development of ICT familiarity and skills rather than any wider educational or employment changes. For example, in the case of Mr Griffiths (a 52-year-old steelworker) the drop-in ICT facility at a local adult education centre had acted as an initial point of ICT access and then also provided a supportive environment to learn basic computer skills. With this 'scaffolding' he had then acquired a computer for the home where he was now refining his computer use and skills with the support of his son and wife:

We never have bothered with computers ... but when I was made redundant in the works, I'd been on and on about – now I had the opportunity to buy one. I went to the education centre first, started learning how to use them and took it on further from there.

After the initial introduction to computers in the learning centre Mr Griffiths' son and wife had bought him another computer (with his son setting it up in the home) with his other children also providing peripherals such as printers and scanners. From this family-assisted start Mr Griffiths (and now his wife) were continuing to develop their skills at the centre as well as at home. Similarly, for Mr McKecknie (a 66-year-old retired telephone engineer) public ICT sites had provided a means through which to make initial contact with computers and develop a familiarity with the technology – a decision, as with Mr Griffiths, prompted by a significant life change:

I retired more or less when I was sixty. I didn't even dream of [getting a computer]. I had enough to do. Later this house came up and with some help I was doing that and I wasn't particularly interested in getting [a computer]. Then I had a heart attack and bypass and all that and in that last two and a half years I've decided I need to get with it. I don't know why. I think not knowing something about it worries you! I went to the library and you get half an hour free ... I set up email and I use the web for various things but I've had no immediate goal. I decided after a year that I might as well get one so I got one. Then I started to go to the college to find out how to use it.

As Mr McKecknie explained, having developed the familiarity and skills from the library and college ICT centres, he had recently purchased a computer for his home; 'I think it's because other people have them and I was just fed up of going to the library. It's convenient [at home]'. These last two cases highlight the role of public sites as being ICT usage points as opposed to mere access points, with users being supported by expert staff.

Despite these examples we would not wish to (and indeed on the basis of our data are unable to) suggest that these instances of empowerment through ICT sites use were commonplace. Even when we specifically targeted known users of public provision for interview their engagement with ICT sites was, more often than not, fragmented and less satisfactory than the previous examples. Indeed, a larger proportion of users of public ICT sites in our interview sample had only done so on an occasional basis ($n = 17$). An extreme example was one individual who, despite being part of our targeted 'booster' sample of centre users, had never used a computer in a public site before the day of interview; having been in the library where he was interviewed using the open-access internet facility for half an hour. He 'couldn't see the point' of using ICT in a public site, although 'the reason why I was [in the library at the time of the interview] was because I was giving my motorcycle a service and I had to wait for it to be done so I took some work and did it in the library' (male, 37 years). Whilst this is testament to the flexible access ethos of public ICT provision, it does suggest that many of the users in our interview sample had, at best, 'low levels' of engagement.

The nature of much of this occasional use of public ICT sites could therefore be characterised as haphazard – both in terms of how it was initiated and the form it eventually took. For example serendipity figured highly in the nature and eventual effectiveness of such 'light' engagement. As this man from Ely explained, his experience of making use of one of the drop-in centres on his housing estate had come about through his casual acquaintance with the centre manager: 'It was just the computer class that I went in for. I knew Lee, the guy who runs it ... I knew him from school. So, you know, he said, 'come along, we can perhaps teach you something' (male, 35 years). This brief use of the centre had resulted in the gaining of some ICT skills but very little current use of computers: 'I've learnt how to use computers properly now ... but I've got nothing to use them properly for'. Another interviewee had gained a casual job as a caretaker at a college which had then opened a public ICT centre. Whilst he was working he would 'sit in' on computer courses, 'not officially, but in the course of my duties I would go round – I've got an hour here, well, I'll sit in on this one, which I did' (male, 65 years).

In other cases these occasional uses of public ICT sites had led onto (or reinforced) more extensive use of ICT elsewhere. For example, as this now extensive user of the world wide web recalled, his first encounter with the internet was whilst he was in his local library looking for books relating to his hobby of photography:

> Something happened and I was in the library and the wife said something and I said, 'hang on a minute' and you pay a pound down there, so I paid the pound. Started tapping in what I wanted – I think it was a book I was after – and I got straight into it and I didn't need the book then, I got straight into the information.

> (male, 63 years)

However, in all these cases any use of public ICT was not sustained – with regular use eventually coming through the acquisition of a home computer. As this 55-year-old computer user explained, although initially using the computers in his local library he had not made use of them since due to the 'private' nature of his continued computer use and the 'public' nature of the library setting:

> The last time I was in there with the computers, there were more old-age pensioners on them than young people. So I mean, fine, you know, good luck to them. I don't think I'd want to use that public one anymore. I'd be frightened I'd swear because I'd done something wrong. And they'd say, 'what a rude bastard he is.'

Interviewer: I suppose it's quite quiet down there?

Yeah. I mean, my neighbours are used to me effing and blinding, I normally play with it out in the kitchen. Sue next door just says, 'Mike's on the computer again'.

(male, 55 years)

This problem of the 'public' nature of ICT use in public sites was also raised by another interviewee who had used her local library's computers on one solitary and unsuccessful occasion:

I was going to go there to learn. And I suppose because I wasn't au fait with it – and none of us like to look foolish – and because the librarian had made a bit of a hiccup with the timing when I was due to have my hour, all those things conspired that I sort of, covered with blushes, I left. I thought, no, this is just too embarrassing for words. It's too personal. I don't want to parade my ignorance to all and sundry. But then our library, [the computers are] out in the open – everybody knows what you're doing. If they were in a little cubicle that might be slightly different.

(female, 65 years)

Other 'once-only' users had specific complaints or problems with the unsatisfactory nature of provision or organisation of provision at their local ICT centres – as one woman complained of her local centre, 'I can't get any satisfaction … they don't know where they are at the moment' (female, 57 years). These reasons notwithstanding the importance of individual motivation to make sustained use of public sites was an over-riding feature of these interviews. This was perhaps most starkly illustrated by Mrs Turner, a college lecturer who had been offered the opportunity to sit in on a class taught in her college during her working day to learn computer programming. She had previously learnt computer programming but needed to 'refresh' her knowledge for a course she was co-ordinating. In terms of having a direct need to learn, a highly convenient course on offer (in terms of time, location and cost) and a continuous history of learning very successfully, it would be reasonable to expect this interviewee to participate. However, as she explains, her lack of motivation (expressed in terms of 'time') overrode these favourable circumstances:

I really do want to know what the [students] are doing. At the moment I just say 'well I might be module leader but I don't know much about it'. But actually it would be quite nice for me to know [the programming language] too. And I still want to do a course that we do in software at work. We do a quite simple introduction to programming and it might remind me of all of my programming.

Interviewer: Why haven't you done it yet?

Time. Time. And I guess priority and other things that have become really important. But it is time because if I'm going to do it I'm going to do it properly. I don't want to start it and give up because I haven't got time.

Interviewer: But these software courses, they're at your college are they?

Yeah. I'd do something at work. I'd just take one of the modules that our students do … and just sit in the back and take the class. I want to do it, I want to have it to take classes and do them. I wouldn't do an evening class. I work at the place so why not just pick up something that's there?

Interviewer: So why haven't you done it yet?

It's purely motivation. Anyone who wants access even around here, you can get computer access. Most people can afford their own anyway [but most] just footle around the edges like me … but as long as you're happy.

Why are adults not making more use of public ICT sites?

As the majority of our respondents had not made use of public ICT sites we now go on to to examine the experiences and perceptions of the non-users. Again it is possible to consider these in-depth interviewees in terms of two broad (albeit not homogeneous) groups. On one hand were non-users who were otherwise relatively extensive users of ICT (n = 34) and on the other hand were individuals who made little or no use of ICTs in their everyday lives (n = 42).

Many interviewees who were relatively experienced users of ICT saw public provision as being too 'low-tech' for their needs. An illustrative example of this perception was Jane Timmins, a businesswoman in her thirties who used different ICTs in most aspects of her day-to-day life. As she explained, being able to use the internet on an 'anytime, anyplace' basis in public spaces was very appealing – but her level of use was such that she was seeking ways to plug herself into her own ICT networks using her own technologies. Using current forms of public ICT provision would therefore be an anathema to her 'pervasive' use of technology:

I think Wi-Fi's a great idea. I've got a laptop as well as a desktop, so I take my laptop with me sometimes and I would love to use Wi-Fi to surf in Starbucks or something, if that was available. So I think in terms of public access, I'm looking to be able to plug me into the net in a public place rather than using other people's equipment.

Of interest from a sociological point of view is how this reasoning echoes the theoretical perspective of actor-network theory in depicting the extensive ICT user developing a fluid network of human and technical elements, with the user being an integral part of the technical network and the network being an integral part of the user (e.g. Bijker *et al.* 1987; Law and Hassard 1999). In Jane's case using public ICT facilities would not afford use of *her* network. For other interviewees who were making extensive use of technology, public ICT sites were viewed as something which would only be used if in new or unfamiliar locations where their usual technological networks were unavailable – although the notion of being outside of one's comfort zone had relatively modest boundaries:

> I wouldn't usually use the library or the internet café, unless I was in a foreign town. I say a foreign town, Bristol even [neighbouring city only 10 miles away]. If I was in Bristol and I knew I needed to check my email, then I might consider it then.
>
> (female, 38 years)

This notion of fulfilling all ICT needs at home is an obvious one when it comes to explaining the selective appeal of public ICT sites – reflecting Facer *et al.*'s (2001) observation that for most ICT users 'home is where the hardware is'. As Jane Timmins reasoned, 'if I've got something I need to do, I come home and do it'.

This preference for home ICT facilities was also manifest in some interviewees' negative perceptions of public ICT provision as being 'second best' and of a generally inferior quality. Such negative perceptions were not solely concerned with the quality of equipment provision in public ICT sites. For some experienced ICT users, public ICT sites of all types were seen simply as something which was 'not for the likes of them'. Perceptions of the 'exclusive' intended clientele for public ICT sites was evident with interviewees in some of the most affluent and some of the most deprived research areas. As an interviewee in Chew Valley argued in response to a question regarding what community ICT facilities were available in the area, 'this is not a community!' – her inference, perhaps, being that 'community' was a euphemism for a deprived area:

> [This] is one of these strange villages where a large number of people go and work in [the nearby city] and their kids go to school at private schools. It's a commuter village; it's not a community village in the same sense.
>
> *Interviewer: So you're not down the cybercafé every Saturday!*
>
> Cybercafé! I wouldn't know where to look for one! ... This is the sort of area where people who want it can well afford to provide their own internet access.
>
> (female, 50 years)

Interestingly, this notion of public ICT sites' exclusivity was also raised by individuals in some of the more deprived areas in our study. As this interviewee on the Ely estate explained, much of the 'community' ICT provision on the estate was not intended to be used by all members of the community:

> They have computers in the Methodist Church.
>
> *Interviewer: Do people use them?*
>
> Yeah [but] it's mainly just the congregation, because they're people they trust. Ely's a very good place, but you've got to be careful who you watch – who you trust and things like that, you know. I've lived in Ely most, well all my life. Like I say, I wouldn't tell 'em, I wouldn't tell my neighbours that I've got a computer because they're into drugs and stuff like that.
>
> (male, 35 years)

Whilst public ICT sites are perhaps not intended for extensive use by such 'established' ICT users their views do begin to uncover the barriers faced by public ICT sites in facilitating 'universal' use of ICT. We next consider the interview data collected from some of the key target groups of the government's current public ICT drive in the form of individuals who were making little or no use of ICT in any form. Here the social context of public ICT provision – in particular people's perceptions of the social context – was a recurring theme. As this 28-year-old unemployed man explained, the municipal sites such as libraries, colleges and (in his experience) job centres where ICT sites were placed were often seen as alien and unfriendly contexts in which to spend time – let alone use a computer:

> It's not practical to those who need them … all these places where you have access … these places are not friendly environments and not places you want to spend time – like job centres where you don't want to be in there longer than you have to. I wouldn't want to spend any time in there at all, just do what you got to do and get out.
>
> (male, 35 years)

As well as the social context proving a barrier, the nature of the ICT-based activities taking place in such centres was also seen as prohibitive. As this recently retired woman explained with regard to the computer provision in her local library, 'I think those computers are there for people who already know how to use them, who've got access to computers at work and, you know, who just need a bit of extra access' (female, 65 years). This notion of public ICT sites being *only* for proficient ICT users was a recurrent theme, as this younger woman elaborated in two separate interview extracts:

You see posters round town for 'computers don't byte' and stuff like that – and that doesn't really reassure you, 'cause you kind of think it's going to be full of people who know what they're doing in there and you don't want to walk in there and be stuck out … most people don't want to stick out at all, do they?

[…]

I suppose it's not that well advertised I don't think. I mean, obviously it is stuck right in the middle of town, but I think people see that it's a computer centre and if they haven't ever used a computer … like the internet café, we wouldn't never go in there, because, to me, it's probably for, you know, like people who are really into the internet and know exactly what they're doing. And I feel a bit of an idiot walking in and sitting at a computer and expecting somebody to come and help me, I suppose, so. I think that is what puts a lot of people off doing computer courses – or any kind of courses, really.

(female, 22 years)

Issues of perception aside, it is also worth considering the key argument for public ICT sites in terms of their ability to overcome existing barriers to accessing ICT (such as cost, lack of ownership, distance, lack of skills) and therefore facilitate 'universal' access to ICT amongst those individuals otherwise without access. There were a few complaints in this respect regarding the lack of accessibility of 'local' ICT sites; for example:

I have got a car but I don't like driving at night as most of the classes are towards the evening and through the forest as well … They've got some beautiful schools over here where they could do it but they don't, it's an absolute waste.

(female, 57 years)

Interviewer: Have you thought about going down to the library and using it there?

I have done, I asked them some time ago and they were charging £2 for half an hour but it's free now but I haven't got a lot of time during the day.

(male, 64 years)

There were also a few examples of a lack of knowledge of what public provision was actually available – as in this example:

I think it's a pity the government can't allow a grant where [a computer] can be put in like a community centre where [people] could have access to it. You can go on the internet, there must be a way of doing it financially and I'm sure people would use them. It's a pity there aren't the resources.

(female, 52 years)

More prominent, however, was the issue of people's lack of reason and motivation to make use of ICT (and it follows public ICT sites). As with the survey data, many of our interviewees recognised the potential to use ICT in public sites if they wished – but reasoned that they had no interest, impetus or need to do so. This rationale was given by interviewees across a range of ages and social backgrounds, as in the case of this solicitor:

> Well, there's the library has it, obviously, and it runs courses every so often. They're very good; they'll help you if you want to go on the net there. There's a youth club just here on the main thingy and that has a good computer bank ...

Interviewer: Is that available to anybody or just members of the youth club?

> I don't know quite what they've done with it recently, whether they have opened it up. The idea was that it would be opened up, because the youth club only operates in the evenings (from four or five o'clock till eight or something). But the idea was that it was going to be opened up and could be used during the day as well ... A lot of these are outreaches from the community centre, the leisure centre – and they are good, because they involve a lot of people and lots of people have access to them.

Interviewer: You've not gone down to the public access computers in the library?

> No. I think we have thought about it a few times, that we ought to go and do it, but we haven't.

Interviewer: 'We' at work or 'we' at home?

> Either. Me, myself – I mean, the family don't need it because they're alright with computers. I thought that I should go down there. And we thought, from the office, we'd go down and do these things, but we never did. Even within the office, it's just time ... you could do this – you can just do the modules, but I never have time ... I mean, I personally don't think they're that difficult to learn.

> (female, 48 years)

Similarly, as this retired man argued, despite making frequent use of the library and receiving encouragement from friends who were already computer users, he had no interest or need to begin using computers:

> I'm up in the library three or four days a week, one way and another, looking up things. I use the libraries a lot. If not to take books out, to look things up.

Using the internet is the modern thing to do [but] I have never actually asked them to find something on the internet.

Interviewer: Have you ever been tempted to have a go?

No, I have just no interest. I go to friends' homes and they say, 'come and look at this' and they fiddle about a bit and it seems to take ages to get onto it and then they press the wrong button or something and it's not engaged. No, I'm afraid I haven't. I've got no need, no intention or need to use it. If I was working I probably would. So you're stuck now, you haven't got anything else to ask, have you!

(male, 72 years)

As another non-ICT-user argued, if (and when) he did decide to begin to use computers then he would do so by 'finding his own way' rather than going to a public site:

Interviewer: If you were tempted to have a go on it where do you think you would go, would you go to somewhere like the library?

I suspect we will have one eventually, I'm sure it will come eventually, and if I did we would just sort of find our way on our own sort of thing, you know, I suppose I'd be frightened of making a fool of myself when I tried to use a computer anywhere else. Yes I'm sure my wife will have her way in the end … she usually does.

(male, 63 years)

There was a sense that this widespread lack of need and/or interest to use ICT jarred with the high-profile marketing of public ICT provision. Some respondents were therefore more militant with respect to their lack of take-up of public provision of ICT. As Mr Turner argued with reference to the sponsoring of the popular UK daytime television programme 'This Morning with Richard and Judy' in the early days of the government 'learndirect' initiative:

You can brainwash if you want to cajole some people to do computers. Well, they're not going to make me because Richard and Judy tell me to. If anything that'll put me bloody off, I should think!

(male, 55 years)

Discussion

While it is clear that public ICT provision is a peripheral part of everyday life for the majority of people, public sites were fulfilling valuable functions for some individuals. Whereas it is perhaps too crude to characterise all users in terms of just two groups, there was a distinct split between frequent users and those who had 'dipped' in and out of using public ICT sites. Within our interviews with frequent users there was also a distinct group (usually younger and less well educated) who were using public centres on a sustained basis as a means to return to education and employment – resulting in a substantial use of the centres both for support and credentialisation. On the other hand, we found some individuals (usually older and better educated) who were reliant on the centres to support and 'scaffold' their use of ICT and development of skills and familiarity.

Although public sites were facilitating empowering changes for a minority, to claim that they were *leading* to such changes is to ignore the nature and circumstances of the participants. Frequent users such as Mrs Smith and Sarah Jones were archetypal 'returners' to education and work. They had already made the decision to 'better themselves' ('to get something behind me' as Sarah put it) and were either directed towards public ICT centres or had chosen computer courses as the most high-profile adult education provision on offer. It is highly likely that both these women, having already made the crucial decision to return, would have engaged in other forms of learning or training were it not for public ICT centres situated in their local colleges. Thus, whilst public ICT sites are catering for such learners they cannot be said to be 'creating' returners. For Mrs Smith and Sarah Jones attendance at a centre did not 'lead' to progression onto an educational course, because this was why they were attending the centre in the first place. Moreover, although such 'success stories' are alluring it must be reiterated that they relate only to a minority of our respondents. Indeed, as we have seen in the survey and interview data, the overwhelming majority of individuals taking part in this study did not make use of public ICT sites in these sustained ways and could not see themselves doing so in the near future. This is not to say that respondents did not acknowledge the 'formal' possibility of using ICTs in a public site – rather that in practice they were not doing so. It should be remembered that nearly half the initial survey sample were not using ICT at all, despite the public provision on offer.

So why do public ICT sites appear to be a marginalised source of ICT use for all but a minority of the population? There were some indications from our interview data that the organisation of the centres was proving to be unattractive to potential users. For example, issues of inconvenience of location, limited opening hours and a lack of awareness were raised as problems for some inter-viewees. Preconceptions of centres' limited technological capabilities were an issue for more experienced ICT users, with a 'snobbery' pervading many people's perceptions of 'public' and 'community' ICT access as being second rate and inferior to facilities at home or work. Of course some, if not all, of these issues of perception and inconvenience can be addressed by public ICT sites. These and other organisational problems and issues currently faced by public ICT sites may well prove to be unimportant in the long term. Nevertheless, our data also raised other more fundamental weaknesses to the notion of public ICT provision as it currently stands.

First, the social contexts surrounding ICT centres were seen by many interviewees as inhibiting access and use. For example, the issue of public ICT sites being perceived as accessible and shared community sites is a crucial issue and one in which the current system of provision appears to fall short. As Shearman (1999) reasons, to be effective ICT centres should either be locally owned or deeply involved in the local community. Although sites such as schools, libraries, colleges and museums may be physically located in communities, they were not always seen to be deeply connected with the communities in our study. Indeed, despite the political discourse of public ICT sites being established in pubs, sports clubs, village halls, supermarkets and leisure centres (Williamson 2003), the vast majority of state-sponsored ICT provision in our areas of study (as in the UK as a whole) was located in existing public-sector institutions such as libraries and colleges. This is predicated largely on issues of convenience, e.g. the cost effectiveness of such institutions in terms of utilising existing buildings, resources and staffing, their existing public-service ethos as well as their existing alignment with the discourse of public provision for literacy and knowledge (Liff *et al.* 2002). Whilst these may well be conveniences of organisation and management, there was a sense amongst some interviewees that they were not using facilities because they did not feel 'part' of either the institutions or the ICT-use taking place within them. This mirrors other studies that suggest, for example, that the take-up of adult education courses in colleges or the borrowing of books from public libraries is attractive primarily for certain social groups who are already well versed in such practices (Smith 1999; Gorard and Rees 2002). Thus the 'institutional' barriers which prevent people from entering facilities such as a library or adult education institute are unlikely to disappear merely because a site of 'free' ICT access has been located within them.

Also important was the wider issue of the (ir)relevance of ICTs and ICT-based activities to many individuals – constituting perhaps the most serious impediment to widespread use of public ICT provision amongst those without access elsewhere. It may be that the present 'problem' does not lie with the levels of supply of public ICT access but the levels of demand for ICT use that exist within the UK population. As Williams and Alkalimat (2004) have observed, the establishment of public ICT sites is a growth 'supply-side' industry – often outstripping actual demand for such resources. The establishment of UK Online and similar initiatives has been based on the implicit assumption that ICT use is an inherently useful and desirable activity throughout all sectors of society. As we discussed in Chapter 1, the logic for many authors behind state-subsidised public ICT provision is an imperative towards allowing people to become 'active participants in the digital world' (Servon and Nelson 2001b: 425). Yet the rhetoric of the 'information age' belies the fact that for many people 'actively participating' in the world does not, and will not, involve personal use of ICT. As we have seen from our interview data, having 'no need' or 'no interest' in using computers were powerful rationales amongst the respondents who were not making use of ICT in their day-to-day lives.

The fact that public ICT access may not be needed or desired by individuals is an obvious one. Cook and Smith's (2002) qualitative evaluation of UK Online centres observed that people's use of public ICT sites tends to be very 'goal directed'. Indeed, this was apparent in the highly focused educational and vocational goals of the returners in our study, and also with many of the occasional users and their goal of 'getting started' with ICT. However, it should be of little surprise that many people did not have any ICT-related goals and therefore had not considered making use of a public ICT site. It is proving 'difficult to create circumstances that engage some socially excluded members of the community [with public ICT sites], even when they are explicitly targeted' (Liff *et al.* 2002: 82).

Conclusion

Establishing 'universal' access to ICTs in the sense that all adults have a formal point of access to technology within a reasonable distance is an achievable target for governments to set. Indeed, public ICT facilities were formally provided for citizens to use in many of our research sites. However, stimulating actual use of ICTs within those social groups currently not making use is far more problematic. There was little evidence from our data that this is being achieved by public ICT provision, let alone the stimulation of use of public ICT sites for educational or learning purposes.

Indeed, throughout all our analyses of the data so far, evidence of adults using ICTs for learning and education (be it in a public or private context) has been sporadic. This reflects our intention not to over-privilege instances of ICT-based learning in our data and therefore give a distorted impression of its prevalence. Although a more honest approach, this has left much of the discussion so far within this book centred around why ICT-based learning is *not* taking place within the general population. In the next chapter, however, we make a concerted effort to focus upon the ICT-based learning that was most evident in our study – the process of adults actually learning to use computers. Through this example we attempt to explore how the three social contexts of the home, workplace and public sites come together and how ICTs interact with processes of learning. Although, perhaps, not the most exciting example of ICT-based adult learning, this can at least give us some indication of the issues which may be of importance if some of the basic issues of participation and engagement are overcome in the future.

Chapter 9

The social processes of learning to use computers

Introduction

We now break with our ethos of the previous discussions by giving undue attention to the ICT-based learning which *was* evident in our study. This is an attempt to address some of the project's final questions concerning the nature and form of ICT-based education. In using the example of how people learn to use computers we seek to illustrate how the three social contexts of home, work and community site actually interact with each other in shaping adults' ICT use and learning. The example of learning to use a computer was chosen as it was by far the most prevalent ICT-based educative activity throughout our data. That we found far more sustained evidence of learning *about* computers rather than learning *through* them is telling, but for the time being it offers us the best available example through which to understand the social complexities of what happens when adults actually do learn with technology.

Although learning to use new technologies such as computers is considered to be a fundamental aspect – even an obligation – of citizenship and employment in the information society, the issue of how people learn to use computers has been little researched within social science. Although learning to use a computer is a wide-ranging and often complex task we know little of how different forms of learning contribute to people's eventual use of new technologies and, crucially, how this learning fits into the wider social contexts of people's everyday lives. This chapter highlights the range of formal and informal learning about computers and computer skills that is taking place.

As we discussed in Chapter 1 there has been a noticeable growth in recent years in both academic and practical 'skills' education in ICT with examination and qualifications bodies offering a range of computer-related courses in compulsory and post-compulsory education. Shorter computer skills and 'computer literacy' courses also increasingly dominate adult education and training sectors in developed countries. This formal provision of credentialised computer skills education has been complemented by the concerted efforts of governments and other public bodies to provide a variety of less formal opportunities for people to become competent and confident with computers. When employer-provided ICT training in the workplace is taken into consideration, every adult in a country such as the UK should, in theory, have the opportunity to learn to use a computer via these formal and slightly less formal courses and training – either in the workplace, an educational institution or community site.

Yet we already know from our data that people do not learn to use computers solely via formal education and instruction. The importance of informal learning should not be underestimated. For procedural knowledge like the use of computers, informal learning is at least as important as formal learning provision. As we also discussed in Chapter 1 there are different forms of informal learning involving the individual learner and others. Alongside auto-didacticism, the research which has been carried out into how people use computers suggests that peer-to-peer mentoring involving partners, children, parents, friends, work-colleagues or family are all important elements of developing technological skills and competencies (e.g. Lally 2002; Murdock et al. 1992) – what Giacquita et al. (1993) referred to as the 'social envelope' of computer use. With this in mind, the present chapter now goes on to report on how ICTs (in particular computers and the internet) are learned to be used, by adults. The findings highlight the range of formal and informal learning about computers and computer skills that is taking place and, in particular, can be used to address the following research questions:

- Why are people learning to use computers?
- In what formal and informal ways are people learning to use computers?
- How may the nature and level of learning be socially stratified?

We address these questions via an analysis of our interview and case study data. From a methodological point of view it should be noted that reflecting on the often gradual process of learning to use a computer was a difficult task for many of our interviewees – given its place in many people's lives as an everyday skill which is often accomplished subconsciously. One of our interviewees described it thus: 'I suppose you just build on it over time and I couldn't tell you how I did that, it just happened ... sort of osmosis' (female, 38 years). Nevertheless, as with all our qualitative analysis these data were detailed enough to be able to provide a rich and varied set of accounts of one of the few tangible manifestations of the 'le@rning society' which was apparent in our research.

The role of formal learning in people's learning to use computers

Our earlier analysis of the household survey data showed that the workplace acts as an important site for many people's engagement with computers, with only 25 per cent of computer users having *not* used a computer at work at some point in their lives. Thus, the structure of the workplace in providing learning opportunities to become an extensive computer user was apparent in some of our interviews. As this woman (who had later gone on to set up her own web design company) recalled:

I've got a very logical mind and I love technology. When I worked for Tesco's [UK supermarket chain] one of the first things they did was they put in a computerised bonus system, and they were using the very first IBM PCs and they needed somebody to train the staff to use them, so that was one of my first jobs with computers. They trained me in computers and then I trained the staff on the shop floor.

Interviewer: Did you volunteer for that?

Yeah, it was a vacancy that came up and [I said] 'yes please!'.
(female, 38 years)

These successful instances of work-based formal learning occurred usually when training was seen to be at a 'high-level' and of direct use or benefit to the individual's job. Another interviewee recounted a two-year training package provided to him during his work as an electrical engineer; 'I really enjoyed that because it was really something to do with my work' (male, 31 years). However, for other interviewees, these formal instances of work-based learning to use computers were seen as low-level, coerced and less successful. As a local government employee observed, 'I think they felt obligated to send you' (female, 61 years). In these instances, the notion of taking part in ICT training at work 'because you had to' prevailed and was perceived more as a chore (both on the part of the employer *and* employee) rather than as a beneficial learning opportunity:

As soon as they brought the system in, we all had probably about an hour's training … I have to say in [the bank]'s defence – they did keep us all up to date with their particular systems … Well, it was in their interest, wasn't it? … When they were bringing in a new program, we all had to toddle off and do a little test and get a little certificate to say we'd done it.
(female, 51 years)

On many occasions this formal learning was based around a 'cascade' model of work-based training. As an office worker explained, 'through the company only so many are actually allowed to be trained and they have to pass the information down to us' (female, 55 years). Formal learning in the workplace also appeared to be inherently entwined with informal follow-up learning at work *and* home – with workers expected to reinforce computer skills on the basis of any formal training received:

It was always very gradual and if there was ever a new part of the system I had to use then a senior would show you, it wasn't structured training, just half an hour where you would be shown. [I learned by] trial and error, it was just that someone in the office would know a bit and then they would show you a bit; it was just picking it up. My husband is a systems analyst and he has been on several Microsoft training courses and he has several training manuals so I used to take them to work.

(female, 36 years)

Aside from the workplace, the role of educational institutions in formally introducing people to ICT and shaping their subsequent use was also evident during the interviews – although less prominent than might be expected given the UK government's emphasis on education technology throughout the 1980s and 1990s. School and university were, for some respondents, seen as key sites where computers were first successfully encountered. As a currently unemployed graduate recalled, 'we used the computer *properly* when I went to university' (male, 39 years). Nevertheless, the value of ICT use in schools and university settings was often recalled in ambivalent or even negative terms, with many students not receiving sustained and effective exposure to computer resources and teaching. The perceived ineffectiveness of school and university ICT provision was not confined to older interviewees – as this recent school-leaver recalled:

We had a computer course in school, I was always way behind then, instead of listening I was playing the 'worm game' [on the computer] … it was either doing that or RE [religious education] … Sometimes we did stuff but nothing I remember.

(female, 22 years)

Community settings such as adult education centres, colleges and libraries were also cited as sources of formal learning about computers by a few interviewees. As with the workplace the nature of this formal training (as well as individuals' motivation for taking it) was often not straightforward. For example, an unexpected feature which emerged from some of our interviews with older adults was how they had chosen as parents to learn to use a computer from formal courses in order to then informally teach their children – reflecting the era of home computing and self-programming of computers in the 1980s. In the following quotation, the mother of now grown children relates how she had enrolled on two formal computer courses in order to 'cascade' the knowledge to her children and husband. This formal garnering of skills proved to be a selfless task – with her personal use of computers soon ceasing once other members of the family had benefited:

I bought my children a computer … One of us [husband and wife] had to learn how to use it so we could teach the kids. So I did a beginners course on computers. Once the children did it then I didn't need to use the computer. Now they are all qualified on computers, I wouldn't like to use one.

(female, 57 years)

As we discussed in Chapter 8, the incidence of interviewees taking formal courses in public and community settings was low, although for those individuals who had taken courses the formal provision had sometimes acted as an initial impetus to becoming a more regular computer user. Some interviewees recounted how taking ICT skills courses in colleges and adult education centres had given them a basic familiarity before purchasing a computer for home – what Lehtonen and Sundell (2004) categorise as 'anticipatory self-education'. Other interviewees recounted taking ICT courses as a strategic means of developing job skills. For example, as this women who had self-financed her way through a postgraduate course in computer programming (despite having no first degree) explained, taking formal computer courses allowed for an ongoing repositioning in the labour market:

> My father was doing some work with a firm involved with consultancy and there was a suggestion that there might be some use for somebody who had Excel [spreadsheet] skills during the course of their consultancy so that was the reason I did the courses … it was more a vague theory.
>
> (female, 33 years)

Nevertheless, in many other cases, taking a formal computer course was less effective and transformatory. As we saw in the last chapter, we came across 'serial takers' of formal computer skills courses – enrolling on courses more for their own sake rather then developing and using computer skills to use elsewhere. As this retired, female 'low user' (whose main use of computers took place via courses rather than at home) explained:

> I've been signing on for computer courses in Radstock intermittently. When they first started off in about, oh God, the early '90s … I've tried so many times … I think a key point is that you have to pay for it. Because if you don't, you don't take it seriously … you just race through this disk and it's all very nice, but you just get used to taking tests, you don't actually learn what you're doing until you actually have to use it. So this year I've paid £125 and that gives me five modules … [but] I can see Christmas is going to be with me, so I probably won't carry on with that …
>
> (female, 64 years)

In many of these instances, it appeared that learning to use a computer via formal courses in community sites was not, on its own, a route to using ICT in everyday life. More important was the informal follow-up where skills and knowledge were developed and reinforced. As this health therapist explained, taking a course in an adult education centre had then led to the development of a sustained informal learning network:

> Not many people that I knew, at the time, knew an awful lot about computers … a friend of mine told me about the course and then I went, then I booked in with her and my sister and we all went…

Interviewer: Oh right, as a group?

Yeah … Lindsay, my friend, has now gone on to do a diploma in something to do with computers. So [now] I tend to sort of ring her if I've got a problem.

(female, 36 years)

The role of informal learning in people's learning to use computers

For the large majority of interviewees, learning to use a computer (as with using a computer) was often a solitary activity. The importance of informal self-education was therefore reflected in many of our interviews. For a few respondents – in all but one case male – learning to use a computer was expressed as an informal learning 'project' akin to other discrete learning projects such as car maintenance or animal husbandry. As this postal worker who was now teaching himself to program PCs and build websites explained, this sustained learning interest often began in childhood with computer games:

I just wanted to find out how the things worked; just wanted to kind of be able to do things on it, so I just taught myself, and took it from there really. … When I was young my dad bought us a ZX Spectrum, one of the early 16K ones, I guess 'cause he wanted us to find out about computers and keep up to date himself … and I just learnt how to program from there.

(male, 33 years)

These processes of self-education were, more often that not, expressed in the interviews in more mundane, haphazard terms: 'generally you can sort of muddle through or see what's happening' (female, 54 years); 'I learned by playing about with it and getting in a mess with it and then learning how to put it right' (male, 31 years); 'I taught myself, trial and error, I got books from the library and just got on with it, stuck it in and see what happens' (male, 64 years). Yet in all these examples of 'solitary' learning to use a computer, learning was more often than not sustained by others, at least on an occasional basis. In this way we would not claim that informal learning to use a computer takes place in complete isolation. For example, we should not overlook the considerable influence of the family and household in shaping individuals' informal learning. For many of our interviewees, learning how to use a computer was primarily achieved from a combination of self-learning, experimentation at home and learning from others. It was clear that most interviewees relied on one or more 'significant others' who acted in an initial mentoring capacity. For some older respondents this role was taken by grandchildren – although this did not necessarily lead to sustained use: 'it's ridiculous really, because my granddaughter taught me how to do it [send email]. I mean I can type but the actually sending it out. The only way I'm confident is that she showed me' (female, 62 years).

Adult family and friends were also important (if irregular) elements in many of our interviewees' adoption of ICT. There were many examples of computer expertise being acquired through the extended family and social networks. This informal acquisition of computer knowledge and advice from others was a recurring theme through our interviews with all age groups but, in the case of older adults, was predominantly initiated and executed by grown-up children:

> [My son has] done about five years on computers at college and got a certificate for it. He's got two or three of his mates that are into computers. They sit down and chat and go over the pub and they sit down and talk computers. And he gets all the information. He keeps up with it, you know. There was a virus the other week and he just phoned me up and said, 'Delete it quick, there's a virus going round'. So I did, you know.
>
> *Interviewer: So is it just your son then, or is there anyone else…*
>
> Nobody round here is really into computers … the bloke next door, he's got a computer … but I never bother them, I just ring my son up, any problems.
>
> (male, 63 years)

For married couples the partner or spouse was sometimes cited as a source of support and learning – often as a result of an 'overspill' of work-related skills or specific competencies. Although our survey data suggested that women were more likely than men to rely on partners this was not always the case, as this man explained with regard to learning to use the Microsoft Word package: 'My wife uses word processing, that's her job. She's a legal secretary. So my wife, she taught me the word processing part of it' (male, 35 years). However, it is important to note that this family-directed informal learning was not necessarily effective or indeed empowering for the individuals concerned. As this retired woman explained:

> My son came in with the computer and said 'why don't you take this up, mother', plonked a computer in there but he's no good as a teacher, no patience … oh my god, he's terrible at teaching, he loses his temper because he doesn't realise that all these things are new to me.
>
> (female, 57 years)

On occasion, using a computer with significant others seemed not to lead to learning or self-development but a reliance on others to use a computer 'by proxy'. Stewart (2002) has observed how the social division of labour means that some people do not need to 'adopt' computers fully but can, nevertheless, experience many of the benefits of computers through others. One of our interviewees, who talked very confidently about the computer during the interview was, in effect, a non-user. When in employment she had never used a computer for anything but word processing. Now, having retired, her husband operated the computer and left her only to type (as she described it, her husband 'does the click-click' – i.e. operates the computer mouse) – a model of engagement mirroring her earlier use of computers in the workplace (female, 65 years).

A noticeable feature of the interview data was the general absence of non-family members in people's informal learning (a trend also reiterated by our earlier survey findings). Friends of the family were occasionally cited as a conduit into buying a computer and initially learning how to set it up but there were few examples of networks of friends and associates with expertise in different areas of computing being developed and sustained beyond the initial orientation period. One example, but atypical of our interview sample, of the extended building of 'technological capital' can be seen with this woman who was an extensive computer user and had developed rich sources of expertise which she could call upon when needed:

> [I've learnt] through other people who were interested ... there are people who I know through my previous job where I had a hardware engineer, and if I wanted now to set up a web server, I would email Steve, because I know Steve would be good at that and he'd give me the right information. And if it was something that I hadn't done before, I'd probably look online; I do research online ... I would use newsgroups. And I would ring someone. I would phone the local college ... I suppose it sounds weird, but there are pockets of expertise around.
>
> (female, 38 years)

As this example illustrates, the workplace was a prominent site of informal learning for many people; in particular through the process of learning to use a computer for a job via an 'informal apprenticeship' from others. As one retired interviewee explained about his job in a milk pasteurising plant, picking up computer skills 'as you went along' was a key source of learning; 'you just picked it up through the bloke who was there that was [next to] you, you know. He just told you what to do' (male, 63 years). This process of 'sitting-with-Nellie' (Overwien 2000) was highlighted by many interviewees, from those in professional and managerial jobs to those employed in manual professions – as this university lecturer explained:

> You sit at a computer, in a shared office, and ask 'how does this work?' – and somebody else would show me how it works. It's wrong to only learn when you've got a problem. In theory, you should go on a training course to learn how to do it all. But in practice, I haven't.
>
> (female, 50 years)

Whilst in some instances co-workers also acted as sources of help when using a computer at home (with some interviewees able to draw on work-based advice and technical support for home-related problems), there was also a tendency for some people to use the home as a site of work-related learning about computers. Throughout our interviews there was a notable pattern of 'home practising' of using ICT for the workplace. Indeed, this informal learning to use ICT at home was a crucial but submerged part of what Cranmer (2002: 3) refers to as the 'digital overflow' – i.e. 'the overflow of salaried work into the home'. As this engineer explained:

> I bought [a home computer] for myself to play on and I felt that, if I had one at home, I could at least try and bring myself up to the expected level that the company would expect of someone in the position that I was, you know.
>
> (male, 36 years)

Discussion

The data in this chapter confirm that learning to use a computer is a complex, gradual process which takes place over the life-course. Although sometimes punctuated and stimulated by formal learning, most computer learning is informal, fragmented and specific to the individual's context of computer use at the time. Learning to use a computer was often self-directed yet the (frequently sporadic) support of 'significant others' was crucial. Also clear from our data was that learning to use a computer is part of how people use computers rather than a stimulus towards further use. As such our data provide some interesting examples of the rhetoric and realities of the le@rning society thesis as well as the ways in which ICT-based learning works out in practice.

For example, the human-capital led assumption that an individual will develop their ICT skills to gain future employment was only partially supported. Some respondents had indeed enrolled on computer courses, bought and taught themselves 'computers' all with a view to increasing their employability. Wheelock (1992: 104–5), for example, highlights the 'entrepreneurial cultural' notion of learning to use a computer in order to develop occupational skills and employment potential – learning to use a computer 'for [new] work' instead of 'because of [existing] work'. More prevalent, however, was the range of learning taking place in the home by workers trying to 'up-skill' themselves for their present employment. These different forms of learning in different sites give lie to the contention that the penetration of ICTs into an increasingly broad range of work areas is redefining and blurring the traditional boundaries of 'work'.

People's reasons for engaging with computer learning were not always straight-forward and, therefore, are not easily catered for by public provision. As with all learning, learning to use a computer has both an 'intrinsic-value' aspect (functioning as an end in itself) and an 'instrumental-value' aspect (functioning in terms of a means for the self-realisation of goals) (Zborovskii and Shuklina 2001). Throughout our data the most common form of learning was instrumental-value learning – where individuals learnt to use computers for specific purposes. These ranged from leisure pursuits such as digitising family photograph collections through to learning ICT skills to 'cascade' down to family members. The reliance on informal learning in the workplace was also highlighted by respondents; in particular the process of learning to use a computer for a job via an 'informal apprenticeship'. Of course, learning-in-work in this fashion has long been highlighted as a key site of adults' learning activity (e.g. Brown and Duguid 1991); however, our study also highlights the amount of learning-*for*-work which takes place primarily in the home and, to a lesser extent, educational institutions and community sites.

Whilst 'instrumental-value' learning appeared to be effective, for many individuals learning to use a computer was self-referential and of 'intrinsic-value' – i.e. the 'computer itself was a goal' (Aune 1996: 92). Thus, individuals who had bought a computer, been given a computer and/or enrolled on a computer skills course merely to 'get up to speed' with computers were prominent in our sample. For some people there was simply a 'desire for the new' (Campbell 1992: 44), as with modern consumer behaviour in general. In many ways, unlike other consumer electronic durables, computers are essentially self-referential (Haddon 1988) in as much as the practical need that is being met by their purchase and use is not fully clear over and above their leisure function. Thus, as with Wheelock's (1992) study of home computer use, we found many examples of people learning to use computers with good intentions but with little actual eventual use and/or utility. Other individuals did not have the opportunity to use ICT in their work and were therefore learning to use ICT due to the *underemployment* of their skills and knowledge in the workplace (see also Sawchuk 2003). For many 'underemployed' individuals, therefore, 'the need for self-education may frequently be shaped in spite of rather than thanks to the characteristics of his [*sic*] professional labour' (Zborovskii and Shuklina 2001: 67).

Reiterating a recurrent theme in our data, the majority of learning to use computers was informal rather than formal. Thus, most learning about computers in our study was garnered from a range of different forms of informal learning involving both the individual learner and others. For example, even when individuals had engaged in formal computer education, learning to use a computer was not about formal learning *per se* but how the formal learning was informally used, followed up, practised and refined. Despite a few of our respondents learning informally via chatrooms, bulletin boards, helplines and other 'online' contacts, it was noticeable how the majority of this informal learning took more 'traditional' routes, such as books and word-of-mouth from extended family networks. This preference for the informal was especially apparent in the importance of 'warm experts' in our interviewees' development of computer use – i.e. people not necessarily with formal technological expertise who nonetheless act as competent mediators between the technology and the lay-person (Bakardjieva 2001). In his study of Canadian blue-collar workers, Sawchuk (2003: 98) also identified the importance of informal 'computer learning networks' between peers, acting as key 'resources of collective learning'. These warm experts are important in having both the specialised knowledge and skills needed to scaffold and support the novices' use of technology.

Many of our interviewees had developed networks of computer support which were sporadically drawn on – but interestingly these were more often than not relatively narrow networks of extended family. Indeed, the centrality of friends and acquaintances to sustaining computer use via the exchanging of software and swapping of information and anecdotes was not as immediately apparent as it has been reported previously (e.g. Murdock *et al.* 1992). Instead our study was nearer to Wheelock's (1992: 97) observation of 'families adopt[ing] personal computers [with] computer usage differ[ing] between the household ... [and] spread[ing] between family members within the household'. Crucially, our own data suggested that this form of learning with extended family members was especially important for those respondents who were older, in lower social groups and with lower educational backgrounds.

It is also clear from our data that learning to use a computer is not a 'one-off' event and that a longer perspective on how people learn to use computers is crucial. People accumulate skills over the life-course – building upon and some-times losing previously learnt competencies and, of course, facing the challenge of changing hardware and software protocols – with their ability to use a computer intrinsically linked with what they are using a computer for at the present. In this respect the data in this chapter illustrate our earlier contention that individuals live technological 'careers' mediated by local contexts of individual and shared technology use. Our data also suggest that learning to use a computer – both formally and informally, as described above – remains a socially stratified activity. Whilst it was striking that people did not learn the majority of their skills and competencies through formal educational provision, when it did take place engagement in formal courses was more commonly reported by younger respondents from higher socio-economic groups and with more advanced educational backgrounds. This suggests that rather than specifically benefiting those social groups who need it most, formal computer education is primarily benefiting those social groups which were already proficient rather than widening the skills base to social groups who were not previously skilled. These inequalities were apparent in the nature of people's learning, how sustained it was and the nature of the support and learning relationships that surrounded their learning. ICT-based knowledge is, of course, as socially conditioned as any other form of knowledge (Mannheim 1985) and it seems that learning *about* computers is not a neutral commodity that everyone can equally access.

This chapter also re-emphasises the inherent link between how people learn to use computers and what they use computers for. Of course these issues include established factors such as, for example, economic circumstance or age. Thus these factors shape not only whether people acquire and make use of computers but how they learn about them. For example, in terms of the unequal relations of power in people's learning relationships it was noticeable from our interviews that women tended to find themselves in a subservient role in the social learning of computer skills. Indeed, for women interviewees more than the men, learning to use ICT was heavily structured by the institutional contexts in which they found themselves (Neice 1998). Throughout our interviews this was especially noticeable in terms of the 'technical intermediation' of institutions such as the workplace, the school and the home constituting structural circumstances which prevented them from otherwise making use of ICTs which could be considered relevant and useful to their lives. As in Chapter 6 we saw how the complexity of familial relationships and household structures were crucial to understanding some female users' engagement with computers – especially in terms of ownership and control of computers in the home. Conversely, as Lehtonen and Sundell (2004: 15) found, 'this learning [to use a computer] is clearly seen as a virtue, even sometimes as a duty, in particular for men'.

Conclusions

In attempting to highlight the complexities of ICT-based learning this chapter has focused on a number of pertinent issues. Successful and effective use of ICT would appear to not merely be about 'having' or 'not having' access to formal technologies and formal technology-learning opportunities, but the scope and intensity of the relationships that people develop with technologies and the nature of what they do with them (Loges and Jung 2001). Moreover, if people have a need to learn via ICT then they may do so – but the need and/or motivation to learn appears to come before (and independently of) the technological opportunity. As we have seen, the boundaries between different contexts of ICT use were often blurred or fully integrated (as was the case for some individuals with regard to work and home). There was also a considerable blurring of formal and informal learning. Indeed, as Hodkinson et al. (2003) argue, 'formal' and 'informal' learning should not be seen as a strict dichotomy, as much formal learning involves a significant amount of informal learning and vice versa. We saw how the primary benefits of formally learning to use computers were often in the informal opportunities for further practice and learning which it offers. Thus, people were collecting or acquiring elements of computer learning from different sources, in different contexts and to different degrees of success. Learning to use a computer can therefore be seen as an ongoing process of 'bricolage' – influenced not only by individuals' material circumstances (i.e. access to ICT, economic means) and their motivational circumstances (i.e. their interest in and utility of using computers) but also by wider mediating social structures.

Chapter 10

Making sense of adult learning in a digital age

Introduction

A number of assumptions about technology, learning and contemporary society have been examined over the past nine chapters. The argument for new technologies supporting an increased choice and diversity of learning opportunities, and thereby enhancing equality of participation, has been at the forefront of our inquiry. We have also sought to test the counter claims that ICTs may contribute to a narrowing of education provision and participation for adults. In broader terms, our research has been able to explore the idea that ICTs are now playing an integral role in people's lives. In this chapter we draw together our findings and use them to expand upon current debates surrounding contemporary society, new technologies and lifelong learning. This process of 'making theoretical sense' of our findings then allows us to construct a set of practical recommendations in Chapter 11.

Revisiting our research questions

Two main areas of inquiry have been prevalent throughout this book. At the beginning of Chapter 3 we phrased them as:

- In what ways is access to ICT in the home, workplace and other community settings contributing to learning amongst adults?
- To what extent is use of ICT interrupting or reinforcing existing patterns of participation in lifelong learning?

Seven chapters later, our responses to these two questions can be summarised briefly as follows. First, access to the main ICTs of computers and the internet is translating into actual use for just over half of the adult population. For these adults, using computers for explicit educative or learning purposes is of secondary interest to more immediate tasks such as producing documents, communicating with family members or searching for information and general knowledge. The most common form of technology-based education taking place in home, workplace and community settings is learning about the computer itself – either through formal ICT courses or self-education. Second, it is clear that ICTs appear to reinforce existing patterns of participation in lifelong learning – primarily benefiting those who are already learners or who would have become learners without the 'intervention' of technology. This continued stratification is most clearly the case for formal education but, as far as we can ascertain, is also true for less formal kinds of learning as well.

In short, the overriding message from our data is that there is little 'special' or 'new' about adult learning in the digital age. As with education in general, ICT-based learning struggles to be part of everyday life. When ICT use (and ICT-based learning) does take place it is as *ad hoc*, pragmatic, messy and mundane as everything else. With this complexity in mind it is also worthwhile to consider the five more refined areas of inquiry which we outlined at the beginning of Chapter 3. Revisiting these specific questions allows us to reach a more specific set of conclusions. These are as follows:

- *What are the established patterns of lifelong learning that can be documented amongst particular adult populations?* Throughout the book we have seen that the learning society continues to be a somewhat divided society. We found a clear delineation between those adults who continued to take part in education during the life-course and those who did not. Over a third of our survey sample had not participated in any formal learning since leaving school. There was evidence of these participation patterns changing over time with younger adults reporting more 'front-loading' of education (i.e. participation in a period of further or higher education once having left school and then sustained non-participation). This was largely at the expense of learning in later life. Although as a result we found more 'transitional' learners amongst younger cohorts, otherwise we found there to be little change in overall patterns of lifelong learning (i.e. those who choose to carry on learning) between age cohorts. Whereas patterns of participation are not completely static they are not being transformed by changes in education policy either. Rather, we found that formal participation in lifelong learning continues to be strongly patterned in relation to a number of key variables. Crucially, the majority of variables which could be used to predict whether or not an individual participated in formal adult education were those characteristics already established by school-leaving age – i.e. sex, parental background and so on. Access to, and use of, ICTs made no difference to the statistical likelihood of someone being a lifelong learner or not.
- *Who, amongst those populations, has access to what forms of ICT within their home, the workplace and wider community sites?* We have also seen that the information society continues to be a divided society. We found that people's level of effective access to most ICTs was heavily patterned by their age, socio-economic background and educational background. Their subsequent level and frequency of ICT use were also patterned along these lines, as well as by area of residence. Other factors such as sex were found to play more

subtle – but equally significant – roles in influencing the quantity and quality of an individual's access to and use of ICTs. Outside the home, provision of ICT access in workplaces and public sites such as libraries, colleges and museums was found to do little to widen ICT use amongst those social groups otherwise without. Most adults expressed a preference for more relaxed sites of access (such as the houses of friends or relations) rather than this public provision.

- *What do adults within those populations use ICT for and how does it fit into their lives more generally?* People's everyday uses of ICTs are generally more 'low-tech' than might be assumed by previous commentators. 'Traditional' technologies such as the television and radio remain the most prevalent ICTs in most adults' lives. Levels of mobile phone ownership are also high but usage of mobile phones or digital TVs as multimedia platforms (e.g. for internet access) is negligible. Computers remain the predominant 'new' ICT in most people's lives. Although 92 per cent of the survey sample reported having some form of access to a computer, only 52 per cent had made use of one during the past twelve months – usually for a limited range of applications such as word processing, emailing, world wide web searching and teaching oneself how to operate the computer. Use of ICTs for engagement in formal education was almost non-existent. However, computers, the world wide web and digital television *were* being used for a range of self-directed learning activities – most notably self-education in the use of the technologies themselves. Many of these educative uses of the computer fitted into and around existing 'offline' practices. Thus where adults were choosing to use computers for formal and informal learning opportunities they tended to have already been existing learners. ICT-based learning was, therefore, found to replicate existing patterns of general educational participation.
- *How do adults learn to use ICTs effectively for formal and informal learning activities?* Adults were found to learn to use, and learn through, ICTs in informal and unstructured ways. Even where adults were purportedly learning formally through ICTs in the workplace or educational institutions, the majority of actual learning (as opposed to instruction) was taking place either on one's own ('working it out for yourself') or through interaction with familiar others (what we referred to as 'warm experts'). Even quite 'high-tech' learning was usually part of a range of technological and non-technological learning methods employed by learners who would use ICTs as only one part of an 'assemblage' of learning sources. Thus, learning from computers was often augmented by books, television programmes and help and advice from others.
- *What are adults actually learning through their engagement with ICT environments?* As suggested above, the biggest single topic that involves learning through the use of technology is how to use the technology itself. This is because many of the people currently being taught to use computers and the internet have no declared need to actually do so. Thus, we found pensioners learning how to produce pie-charts from spreadsheets as part of their computer courses

but then never needing to produce one ever again. One drawback with this emphasis on learning the use of hardware and software it that the detailed knowledge gained is so rapidly obsolete. Other than this, most adults seem to be creating a use for the technology, rather than the technology solving some existing problem or deficit in their lives. This was most obvious in hobby and leisure use, such as examples of adults producing greeting cards.

As stressed at the beginning of the book we have tried to be careful to construct an objective and realistic picture of adult learning in the digital age. We have reached our sometimes bleak conclusions while trying to find generalisable evidence of technology-based learning. In reviewing these findings it should be remembered that we used extremely generous and wide-ranging definitions both of what constituted 'adult learning' and 'ICT use'. We are therefore confident that the modest levels of ICT-based educational activity evident in our findings are not simply a result of asking the 'wrong' questions or of under-reporting. For example, with our survey data we were happy to consider someone who used a computer on a once-a-week basis as a 'frequent' user. Similarly, just one reported episode of health and safety training during a respondent's lifetime would have made them a 'participant' in adult learning. Yet even within these wide-ranging boundaries we still found relatively low levels of people fitting either criterion. Also, our results need to be read with the constant reminder that we conducted a household study so that non-participants in both learning and ICT-use were fully included. Most educational research, and almost all ICT-use research, on the other hand focuses only on learners and users, so routinely excluding from research those already excluded from the le@rning society itself. Widening participation studies are generally conducted by asking existing learners what barriers they faced to participation, and so on.

Having presented our findings, we are left with the task of addressing the perennial social science question of 'so what?'. What can be learnt from one specific study of 1,001 adults in four areas of the UK? What can we now say about adult learning in the digital age which could not have been said before? What implications do our findings have for the learning society and information society debates? In order to address these questions we focus on six broad themes emerging from the preceding empirical chapters which we feel are worthy of further discussion. The remainder of this chapter therefore concentrates on the following topics:

- the apparent inertia of participation in formal adult learning;
- the modest role of technology in everyday life;
- non-engagement as evidence of digital choice rather than digital divide;
- the emerging signs of an informal le@rning society;
- the entwined nature of inequalities and stratification of ICT-based learning;
- the importance of locating learning and technology use in a lifelong context.

The apparent inertia of participation in formal adult learning

It seems from our study that patterns of non-participation in adult education remain as entrenched as ever – whether in terms of learning through ICTs or through other 'traditional' channels of adult learning. It is also apparent that non-participation is not *solely* an issue of an individual's circumstances or means. Our study found non-participation in formal adult learning to exist across all social groups. Although a vague feeling of goodwill towards the idea of being able to learn via technology may exist amongst the adult population, even the well-educated and technology-rich reported very low levels of *actual* participation. As Mossberger *et al.* (2003: 77) concluded ruefully from their survey of North American internet users, 'translating this willingness into the commitment to actually take [online] classes is another matter'.

Yet although doing little to interrupt patterns of educational (non)participation, we found that technology *does* at least appear to be playing a valuable role in adult education by helping some of those who want to be helped. Throughout our survey and interview data we found some individuals who were deeply involved in learning through ICTs. We found older adults using the internet for a wide range of informal learning. We found young mothers who were attempting to restart their school education by returning to colleges and ICT centres to take computer courses. We found adults who were learning new skills for the workplace or for the purpose of 'keeping up with' their children. Throughout our research ICTs were sometimes found to be performing an invaluable role in helping such learners to learn, demonstrating the undoubted educational potential of technology. On this basis alone we would celebrate the role that ICT is playing and argue strongly for its continued presence and promotion in the field of adult education.

Yet these positives must be set against the claims and expectations which surround ICT and education. Despite the ICT-based learning in evidence we found little indication that technology was somehow 'creating' new learners – i.e. acting as a source of encouragement, support or impetus for those who were not already engaged in learning. In terms of the categories developed from our survey data, 'non-participants' were resolutely remaining non-participants in spite of ICT-based learning opportunities on offer to them. In other words, as Wresch (2004: 71) observes, 'non-participants are not suddenly participating in adult education because of the advent of online courses'. The fact that existing learners are using ICT as a means to continue to learn is not evidence of ICT 'causing' learning to take place.

We are therefore led by our data to conclude that many of the impediments to participation in adult education are *not* party to 'fixing' via the application of technology or other interventions. It could well be that those predicting the establishment of a fully participative learning society misunderstand the nature of adults' (non)engagement in education and training. In particular, the prevailing model of educational participation based on human capital theory assumes that adults act as rational egoists, making decisions based on market forces and specifically selecting educational episodes based on how they will improve their employability, earning potential or quality of life. They then weigh these advantages against the costs and act accordingly. The logic of the 'knowledge society' informs us that an individual's investment in education will increase their employability, and thus the outcome of this calculation should be an enthusiasm for participation (Fevre et al. 1999). In some respects, prior research has shown at least part of this to be so (Jenkins et al. 2002). Prior lifelong learning experience is associated with a higher likelihood both of subsequent learning and of employment. However, it is also associated with lower lifelong earnings. This means that, even if human capital theory is correct, people can still claim to be economically rational in refusing to participate. The knowledge society model also presents the ability to use technology as a key to thriving in contemporary society – again suggesting a rational enthusiasm for learning how to use ICTs in order to improve quality of life. Yet our data tell us that large sections of the adult population do not engage in post-compulsory education or training – let alone ICT-based education or training. As far as the human capital model is concerned, there can only be one explanation for this, i.e. that barriers are standing in the way of individuals taking the action that they would select if they had a free choice.

However, this assumption runs counter to the limited role that education actually seems to play in many adults' lives. It also appears to overestimate the importance and significance of technology in people's lives. Our data suggest that the idealised notion of the economically rational individual choosing to learn via ICT ignores the 'goodness-of-fit' and the usefulness of both learning and technology within the social complexities of people's lives. In this way a highly salient reason behind the non-use of technologies for learning is simply the (ir)relevance of formal education to people's lives. Much academic and political interest in education and technology has been based on an implicit assumption that formal learning is an inherently useful and desirable activity for *all* sectors of society. Our data would suggest otherwise; formal education may not really be as practically important to many people as educationalists like to think.

When they are used, ICTs seem to fit around (and be shaped by) the existing patterns of people's lives. In this way, the acquisition of a computer or digital television set is likely to reinforce rather than alter what people do in their lives. This tendency to augment what has gone before suggests that ICTs in themselves will do little to disrupt or radically alter pre-existing inequalities. It is not surprising that we find having access to ICT 'failing' to make people any more likely to participate in education and (re)engage with learning. We can therefore conclude that, at best, ICT increases educational activity amongst those who were already learners rather than widening participation to those who had previously not taken part in formal or informal learning. Whilst this can be seen as a perfectly good use of the technology, it remains the case that the ICT-based learning which is taking place is that which is primarily of benefit to the 'usual suspects' (Selwyn and Gorard 2002), i.e. those who have taken part in adult learning before as opposed to the 'previously uninvolved' (Wresch 2004: 71). In fact, such changes can be seen to be exacerbating existing inequalities as they further highlight the entrenched nature of non-participation for those individuals who continue not to participate:

> Far from reflecting a new freedom, social, economic and cultural changes have merely resulted in new forms of the traditional patterns of inequality in society. Opportunities for new life styles have been generated but access to these is fairly limited. Many are just as disadvantaged as they used to be; in fact, in some ways more so because the ideology of individualisation has attained such a hold over consciousness that they are more likely to blame themselves rather than the system when things go wrong.
>
> (Quicke 1997: 143)

We are not the only researchers to reach this conclusion. The Canadian academic Peter Sawchuk (2003: 646) also characterises a 'them-who-has-gets' pattern of participation in computer-based adult learning. Recent studies of the digital divide in the US highlight a similar situation where 'interest in online education was statistically more likely among the educated, the young, the affluent, and the employed, controlling for other factors' (e.g. Mossberger et al. 2003: 77). In fact the 'usual suspects' conclusion is a phenomenon applicable to most aspects of society. The observation that 'them-who-has-gets' is a perennial criticism of attempts to engineer full participation in most 'beneficial' societal activities. As E. Schattschneider observed of efforts to coerce mass-participation in political activity, 'the flaw in the pluralist heaven is that the heavenly chorus sings with a strong upper-class accent' (1960: 33). In this way, there is little new about the 'new' le@rning society – not even in the nature of its inequalities.

The modest role of technology in everyday life

This failure of ICT to stimulate mass participation in lifelong learning is compounded by the relatively low levels of more general ICT use apparent among the population. By 'relatively low' we refer to the fact that only around half of the adult population were found to be engaging in some way with computers and the internet, as opposed to the more prevalent use of mobile telephones, televisions and radios. While these levels of engagement compare favourably to many other countries, the fact that around half of all adults were not engaging with the internet or a computer runs contrary to the rhetoric of universal access and 'pervasive' computing which underpins the le@rning society model. Indeed a fundamental weakness for those seeking to use these ICTs for adult education is that computer and internet use is not a ubiquitous activity. Moreover, when ICT use does take place it is often sporadic and limited in its scope. Much of the use of computers and the internet found in this study could be said to be decidedly mundane and 'low tech' when compared to the high-tech, high-octane rhetoric of the 'information society'.

The discrepancy between the visions of ICT use promoted by proponents of the information society and our more down-to-earth findings of adults' actual technology use is striking. Although high-users of ICT were apparent in all of our research areas, communities such as Radstock, Ebbw Vale or Ely appeared to be light years away from Lash's (2002) description of culturally regenerated 'live zones' at the centre of information and media 'flows' and populated by the 'informational bourgeoisie' and 'new-media cultural intellectuals'. Of course, by choosing to focus on representative areas of England and Wales it was probably unlikely that we would come across such extreme examples of contemporary techno-culture; after all, as Graham (2004: 16) observes, 'the so-called 'information society' is an increasingly urban society'. It could be that the concepts of the information society and knowledge economy do not travel well outside the 'hotspots' of London, Tokyo and Manhattan.

Indeed, our data should be seen as an empirical rejoinder to the description of 'live zones', 'power users' and extensive ICT use which pervades current discourses of the information society. For example, we spent a great deal of time designing a section of our survey instrument which detailed people's use of the internet on platforms other than a computer – only to find that fewer than 30 respondents out of the 1,001 adults in the sample had *ever* accessed the internet on a mobile telephone, digital television set or games console, and even fewer did so regularly. Contrary to popular depictions of our technology-fixated society, the vast majority of adults were found to be eschewing activities which are much written about and discussed in the literature such as 'blogging', using internet chatrooms or downloading music and films. Instead computers were being used for more routinised and mundane purposes such as writing letters or searching the world wide web for information on products and services. We do not intend this to be a pejorative observation on our respondents but a riposte to those commentators describing the widespread, high-tech information society. On the evidence of our research it seems that the information society discourse is nothing more than a highly exaggerated and almost mythologized vision of life in the early twenty-first century, not in evidence in the majority of people's lives. As Gibson concurs:

> It would be imprudent to suggest that such technologies have permeated everyday lives in the comprehensive fashion optimistically predicted in the mid-1990s by some utopian commentators, software manufacturers and dot.com entrepreneurs.
>
> (Gibson 2003: 244)

Indeed our study highlights how most adults, quite sensibly and pragmatically, will use sophisticated technologies for relatively unsophisticated tasks. Crucially for the purposes of our book, this has obvious implications for levels of engagement in ICT-based adult learning. If computers are, for the large part, integrated in pre-existing individual and household habits, norms and values, one would expect them to reinforce rather than transform people's patterns of (non)learning. As such, the assumption that ICTs can fundamentally change educational and learning patterns and habits is unfounded. Whilst some adults do, of course, learn with computers, many others do not – just as they do not learn via non-technological means either. To expect anything different is to imbue the computer with a social capability far beyond its means.

Non-engagement as evidence of digital choice rather than digital divide

It is clear to us that non-use of ICT cannot be due solely to barriers of cost, time, skills or confidence as is commonly assumed. Indeed, our interview data reflect the complexity, fluidity and ambiguous nature of people's (non)engagement with ICTs. Whilst the opportunities to access and use ICTs are undeniably mediated by wider socio-economic factors such as age, class and gender (see below), we cannot deny individual agency in contributing to the fluidity of people's relationship with technology and with learning. Throughout our data, non-use of ICT for educative purposes or otherwise was mediated by conscious decisions and whims, ongoing 'life-flows' and deliberate changes in people's life circumstances. We would argue that for many individuals *not* using certain technologies is a nuanced decision rather than always just a matter of disadvantage. As Bruland (1995: 144) suggests, '[non-use] could thus be seen as a positive part of a social selection process, not an obstacle to the inevitable march of technological progress'. In particular, making sense of and acting upon the 'meaning' of technology in their everyday lives appeared to lie at the heart of why many of our interviewees were not making use of computers or the internet. To assume that non-use of ICTs is due to the individual concerned being somehow prevented from doing so is to ignore the subtleties of the interactions behind the (non)use of technology.

For example, behind the powerful but nebulous reasons of 'simply having no need' or 'no interest in using a computer', a conclusion which could be drawn from our interview data is that a strong sense of ambivalence existed towards ICTs such as the computer – described by Dutton (2004) as being 'informed but indifferent' towards technology. Whereas psychologists see ambivalence arising from intrapersonal conflict, here we can turn instead to the broader sociological notion of ambivalence arising at the level of social structure when an individual in a particular social relation experiences contradictory demands or norms that cannot be simultaneously expressed in behaviour (Merton 1976; Weingardt 2000). We can identify some of our interviewees' profoundly ambivalent attitudes as reflecting various structural attributes of the information society – in particular where people are surrounded by 'macro' discourses and portrayals of inherently beneficial, empowering and 'magical' new technologies from governments, media and even peers whilst at the same time experiencing a fairly limited utility, usefulness and pleasure from the same technologies from a 'micro' everyday life perspective.

In many cases this sense of ambivalence could stem from the current (ir)relevance of ICT to many adults' lives. As we have already hinted at, much academic and political interest in stimulating universal use of technology is based on an implicit assumption that ICT use is an inherently useful and desirable activity throughout all sectors of society. Thus, for many authors, the logic behind state-subsidised public ICT provision is an imperative 'giving people the information tools they need to participate in the decision-making structures which affect their daily lives. It means helping people use these resources to deal with their everyday problems' (Doctor 1994: 9). Yet this rhetoric belies the fact that for many people 'dealing with everyday problems' does *not* involve personal use of ICT. In trying to understand patterns of (non)engagement with ICTs there is a need to reconsider the 'relative advantage' (Rogers and Shoemaker 1971), 'situational relevance' (Wilson 1973) and 'informational need' (Chew 1994) associated with technology use. Where the impact, meaning and consequences of ICT use are limited for individuals then we cannot expect sustained levels of engagement:

> the concept of the information age, predicated upon technology and the media, deals with the transformation of society. However, without improvements in quality of life there would seem to be little point in adopting online multimedia services.
>
> (Balnaves and Caputi 1997: 92)

That ICT may not be needed or desired by individuals is obvious but surprisingly rarely considered as an explanation. On the basis of our data we would contend that the notion of 'digital choice' needs to be considered seriously when attempting to understand ICT-based education and the le@rning society. With other recent studies suggesting that significant proportions of the adult population are simply choosing not to engage with ICT-based activities (such as e-learning, e-banking or e-voting) because of a lack of interest, motivation, need or usefulness (see World Internet Project 2003; Mathieson 2003; Cabinet Office 2004; Dutton 2004), there is little reason to assume that adult education is any different.

The emerging signs of an informal le@rning society

In highlighting the low levels of formal ICT-based learning apparent in our data we do not wish to completely dismiss the ideal of a le@rning society. As previously mentioned, we found several instances of ICTs being used for learning in our study. In particular, we found many examples of the sustained use of computers, the internet and digital television for informal learning. When any of our three activities of interest (i.e. ICT use, learning or ICT-based learning) were found to occur in our study, there was also a theme of people expressing a preference for the informal and the familiar over the formal and unfamiliar. This was apparent in terms of what people did, how and where they did it and who they did it with.

In terms of how and where people were learning, for example, the preponderance of informal learning in our study confirms the observation by Alan Tough cited at the beginning of the book that informal learning is the 'submerged bulk of the iceberg' of adult education. We know from previous studies that informal learning represents the majority of learning that takes place across the workplace, community and home (Livingstone 2000). Given, as we have just discussed, that ICTs seem to fit around adults' pre-existing habits, norms and values, one would expect computers to be assimilated into pre-existing patterns of informal learning rather than leading to any expansion of 'new' formal engagement with education.

It seemed that much of the ICT-supported informal learning documented in our study took the specific form of 'self education' – a more specific and personalised form of informal learning concerned primarily with the individual's ongoing and burgeoning relationship with knowledge (see Gadamer 2001). It has been argued quite convincingly that the majority of most people's 'real-life' learning comes under the aegis of self-education even though the majority of research on teaching and learning tends to ignore it (Shuklina 2001). For some commentators, therefore, self-education offers the most democratic and potentially empowering form of education of all – unhampered by many of the institutional and situational barriers seen to beset formal education: 'self-education has greater opportunities for individualisation than [formal] education, lacks repressive features (penalties for failure to apply accepted norms, failure to comply with models of behaviour, and so forth), and is more dynamic' (Zborovskii and Shuklina 2001: 68). As Douglas (1992) continues, self-education provides a 'freedom of education' and unrestrained 'non-system' of education which should be positioned at the heart of the 'learning society' model.

The failure by educational technologists and adult educators to acknowledge self-education fully is ironic given that it would seem to fit neatly with the portrayal of the learner-centred model of ICT-based education outlined in Chapter 1. It also goes some way to explaining the mix-and-match 'bricolage' approach to learning with computers displayed by our respondents. As Shuklina has previously noted, traditional methods of self-education persist alongside technology-supported self-education:

At the same time, the older methods of self-education that characterised the traditional mechanisms of the reproduction and transmission of knowledge, are not being withdrawn from mankind's arsenal; rather, they are functioning locally within the framework of particular elements of culture, or else they are reproducing particular types of subcultures.

(Shuklina 2001: 74)

This was evident in our respondents' use of books, cassettes, television programmes and so on which surrounded their use of the internet as a learning tool. Indeed, much of the learning to use a computer as documented in Chapter 9 comes under the aegis of 'self-education' as a specific form of informal learning. Such self-education was nurtured both by the individual and others, with the responsibility for its initiation and cultivation lying primarily with the individual but being achieved with the support (or otherwise) of others (Gadamer 2001).

Our data also highlighted the preference for many adults to turn to familiar social settings (however flawed and restrictive) rather than unfamiliar ones when it came to engaging with technologies and/or engaging with learning. This was evident in people's relationships with their children and getting help and advice about using a computer from them instead of using expert tutors or online helplines. Similarly, our survey highlighted how more people perceived themselves as having access to ICT through the homes of extended family and friends as opposed to through public or community sites. This preference for the familiar and informal was especially evident in terms of how people used and learned to use computers. Although a range of different social contacts was mentioned throughout the interview data, the 'traditional' nature of who was involved in most adults' use of computers was striking. Most sources of help were drawn predominantly from individuals' existing social networks – especially close relatives or, to a lesser extent, the extended family. Although others were often used as important sources of information (especially in the initial stages of acquisition), these different social contacts played vital roles in enabling and, in some cases, regulating individuals' ICT access. Many interviewees had benefited, for example, from the lending, borrowing or dumping of computer resources from others. On the other hand, the ongoing use of this hardware was then often subject to negotiated collaboration (or conflict) with the same others.

Given the 'ordinariness' of adults' (non)use of ICT, the reliance of most adults on close (and sometimes extended) family contacts as opposed to more formal sources of support in the workplace or community is not that surprising. As has been observed before, people are more likely to interact with and seek advice from those they are close to: 'people who know and trust each other are more likely to share personal information. If they have a background of shared experience, they can more easily convey that information, and responses are more likely to be interpreted as supportive' (Resnick 2002: 254). With ICTs such as computers, perhaps more so than other consumer goods, conforming to what is seen to be the societal norm of using ICT can be a deeply personal (and potentially embarrassing) process – better shared with family members rather than friends or work colleagues.

The entwined nature of inequalities and stratification of ICT-based learning

So far we have touched only briefly on the considerable structuration of adults' (non)engagement with ICTs for learning purposes. Although we would argue that individual agency was an important theme from our data, we recognise that the patterns of learning and ICT use in our data were often structured along sometimes subtle but often pervasive and highly unequal lines. There were, as noted before, entrenched inequalities of opportunity and outcome in terms of age, gender, socio-economic status and educational background. Although all were found to be significant influences in their own right these factors were often combined and compounded, making their identification difficult.

This complexity can be seen if we take the example of how 'age' was found to be an important factor in determining people's use of ICTs and engagement with learning. First, our data showed that to conceptualise all older adults as polarised between either being 'can-nots' or highly empowered 'silver surfers' is misleading. Indeed, the construction of the highly resourced, motivated 'silver surfer' using ICTs for a range of high-tech applications was erroneous in all but a few cases. Our survey data showed that a lot of older adults' computer use was basic and mundane. The stereotypical notion of the 'silver surfer' using the internet for banking and finances, shopping and dealings with government agencies was not in evidence. Instead the minority of older adults who were using computers were doing so for word processing, keeping in contact with others and generally teaching themselves about using the computer. Using a computer, as well as being a minority activity amongst older adults, was also highly stratified by gender, age, marital status and educational background. Those older adults who were using computers appeared to conform to the younger, male, educated stereotype which has been associated with computer users over the past two decades. It could be concluded that it was the circumstance of being an older adult which was leading some people to use and some people not to use ICTs. Old age could be seen as a context rather than a cause (see Ocak 2004).

Often the stratification observed in our study was subtle and tied up with wider patterns of reproduction of inequalities. Therefore ICT in itself was part of the wider structuration and stratification of society. This can be seen in the example of gender. Although our survey data were not found to differ significantly between male and female respondents, gendered differences permeated our interview and case study data. For example, many women's engagement with computers appeared to be strongly aligned with the pre-existing micro-politics and power dynamics of the household and family. The Dutch sociologist Lisabet van Zoonen highlights differences between 'deliberative' media cultures in some households where partners negotiate the uses of ICTs, as opposed to 'traditional' media cultures:

> … in which computers and the internet are considered to be the domain of the male partner in the household. He uses them most often, knows most about it and is highly interested in these new technologies. In the most extreme cases, he monopolises the computer and the internet.
>
> (van Zoonen 2002: 17)

Although some evidence of 'deliberative' cultures were apparent in our interview data, many female respondents' use of ICT was obviously compromised by their children dominating 'shared' IT resources. Thus we found that any good intentions or expectations which women may have about using computers once having acquired them, 'ultimately collide with the gendered constraints built into the pre-established territories of the home' (Cassidy 2001: 44). From one perspective, therefore, acquiring a computer but not using it could be seen as just another unequal 'family trade-off' implicit in many women's domestic labour (Abroms and Goldschieder 2002). The gendering of ICT use in the home evident in our data reflects the wider gendering of the home where women sometimes lack either authority or a space of their own and have their emotional, spatial and educational needs placed secondary to those of their partners and children (Mallet 2004).

Alongside age, gender, educational and socio-economic background there were some other salient factors found to underlie use of ICT and engagement with ICT-based learning which are more difficult to account for adequately. In particular the influence of geography and place was evident throughout our data – often in subtle and not immediately obvious ways. In both our multiple regression analyses (pertaining to educational participation and then ICT use), we found evidence of spatial patterning above and beyond the dominant shaping variables of socio-economic status and age. These area effects were not uniform – with 'mixed' patterns of engagement with education and ICT evident within our twelve sites of research. It was apparent from our survey data that these differences could not be explained in homogeneous terms of 'rural' or 'urban' – with all twelve areas in our study displaying markedly different patterns of engagement with education and technology.

In terms of which specific aspects of place and geography may have been contributing to these patterns of (non)engagement with ICTs, our interview data were able to, albeit tentatively, highlight a range of tangible issues. These ranged from obvious issues of technological infrastructure (such as the limited coverage of Wi-Fi and broadband internet connections in different locations) to less obvious issues of local histories and identities (such as the inertia experienced in what residents perceived as 'non-computer' villages or the difficulty of applying ICT skills in a local employment market built around quarrying and mining). These issues chime with the growing recognition that the globalising impact of ICTs may well be one of 'glocalisation' – i.e. the concurrent processes of globalisation *and* localisation. As Will Hutton (2004: 26) argues: 'the internet is less about abolishing distance but more about entrenching the depth and complexity of local social relationships'. Yet in the case of our data these factors could be more accurately seen as individually-held perceptions of place rather than universally experienced barriers or affordances. Even in terms of the physicality of place – such as large hills or small houses – we saw how different individuals presented these issues as having varying influence on their (non)engagement with technology. As with all our discussion so far, the extent to which place impacts on an individual's engagement with ICT is ultimately predicated upon individual motivations and dispositions towards ICT – which in turn are entwined with an individual's socio-economic status, age, gender and other life circumstances.

In this way, place 'certainly matters to those who are disadvantaged ... Because they tend to have a much more 'localised' orientation than the population as a whole' (Green 2001: 1363). It can be argued, for example, that the most disadvantaged individuals in our study tended to be the most 'locally orientated' – for example in their perception of opportunities and services beyond the immediate locality. Indeed, Hasluck and Green (1998) identify two distinct groups: those who could be characterised as being 'very local', and those with wider and less locally-rooted perceptions. Thus we saw that those with more economically, socially and culturally advantaged positions were able to transcend or deal with the barriers to ICT engagement thrown up by place if they so wished (by moving to larger houses for example) as well as taking advantage of affordances (such as paying for home-delivered online shopping). The problem here is that many of the area-related issues highlighted throughout our data are beyond the means of more deprived people to alter. As Burrows and Bradshaw (2001: 1345) reason, 'the fortunes of particular neighbourhoods are as much to do with actions (or inactions) of rich people as they are to do with the actions (or inactions) of poor people'.

A final area of difference in our data was the tentative patterns emerging in terms of ethnic group – with respondents in the 'non-white-British' category appearing to be more likely to engage in ICT-based learning and ICT use than their 'white British' counterparts. This could support the conclusion that ICT is acting as a leveller for these social groups. Unfortunately, given our approach to gaining a representative sample of the population areas, the number of 'non-white British' respondents was too small to draw any firm conclusions from (as well as 'non-white British' being an almost meaninglessly broad category). As such we are only able to suggest the possibility that differences exist here. This is something which has been pointed out as a weakness in our research design in the past – although we would counter that our primary concern is one of actually reflecting the learning and technology use in our areas of study rather than holding a brief for particular social groups or areas of academic interest. Nevertheless, it could be argued, as Warren did of an earlier research project, that:

> the multi-cultural nature of South Wales society is almost totally absent from their studies. The justification for this is that the actual number of minority ethnic subjects in their research is too small. Consequently, they are not able to say anything about learner identity, learner trajectories and race/ethnicity. This must surely raise questions about the usefulness of methodologies that cannot adequately give presence to minority ethnic communities.
>
> (Warren 2004: 108)

The importance of locating learning and technology use in a lifelong context

All these analyses and discussions highlight the importance of approaching adults' use of ICT and participation in education from the perspective of the individual life-course. In terms of lifelong learning, for example, it is clear how educational participation is tied up in the ebb and flow of individuals' present life circumstances. Thus, lifelong learning took place or was indulged in as and when it was needed or could be 'fitted in'. In particular, the implications of our regression analysis of the survey data are profound, providing strong empirical support for the utility of the concept of 'trajectory' in analysing participation in lifelong learning. Not only was there a clear pattern of typical 'trajectories' which effectively encapsulate the complexity of individual education and training biographies, but these analyses also showed that which 'trajectory' an individual takes could be accurately predicted on the basis of characteristics which are known by the time an individual reaches school-leaving age. This does not imply, of course, that people do not have choices, or that life crises have little impact, but rather that, to a large extent, these choices and crises occur within a framework of opportunities, influences and social expectations that are determined independently. At this level of analysis, it is the latter which appear most influential.

It was also clear from our data that people's use and non-use of technologies – and their propensity to use technology for informal and formal learning – was shaped as a highly individualised life-history of 'dipping in and out' of technology use and learning. We saw how technologies have their own histories and careers or artefacts. This was apparent in how computers were recycled between extended family members, moved between the home and workplace or slowly altered to become a son or daughter's computer as opposed to an (ostensibly) inclusive family object. As Haddon and Silverstone (1994: 3) observed, 'these may undergo different patterns of use over time, change location, be the "possession" of different people in the household, or indeed leave a household, being either discarded or passed on to another home'.

Our highlighting of 'histories' and 'careers' of learning and technology use lends weight to the argument outlined in Chapter 2 that people are not simply 'users' or 'non-users' of computers or indeed 'learners' or 'non-learners'. Being a 'computer user' is not a permanent state-of-being and having once learnt to use a computer does not irreversibly make one a computer user for life. Instead, as Graham Murdock (2002) reminds us, the influences behind people's (non)use of ICT are multi-faceted and historical – with individuals living technological 'careers' mediated by 'local' contexts of individual and community technology use. Over their lifetime people can therefore move through different states or levels of technology (non)use depending on their circumstances and context. For example, someone making only spasmodic and limited use of ICT whilst looking after their children can progress onto continuous and comprehensive use of ICT upon returning to work. Conversely, core users can lapse into becoming absolute non-users and *vice versa*. Similarly, in terms of learning we would agree strongly with Cremin's (1976) observation that 'individuals come to educational situations with their own temperaments, histories and purposes, and different individuals will obviously interact with a given configuration of education in different ways and with different outcomes' (cited in Weiland 1995: 99). In other words, people's present state-of-being regarding their use of technology for educational purposes is shaped not only by both their present temperament and motivations but crucially by their life-histories of technology use and education.

Conclusion

In reviewing our study findings for their academic substance we hope to have gone some way towards locating adult learning within the bounded, messy realities of contemporary social life. As such the picture that has emerged from our data is neither as utopian or dystopian as other authors would have it. It is clear that whilst new technologies are augmenting the types of educational activity taking place, ICT-based adult learning remains profoundly shaped by what goes before it. In this way, ICT is not a 'break from the old' as many hope. Instead, as Murdock (2004a) argues, the fashionable notions of post-modernity, the digital revolution, cultural globalisation and so on should not distract from the historical continuities at the heart of the information society.

We have focused much of our discussion on issues of social justice, inequality and (non)engagement. This is because these social issues form the bedrock of the official learning society and information society agendas alongside the more often discussed issue of economic competitiveness. Having tested the notion of ICTs leading to a fairer, democratic learning society we would conclude that much must change if the realities of adult learning in the digital age are ever to near its rhetorical promise. It is to the issue of what should change which we finally turn. Pointing out the inevitable shortcomings of practice and policy when set against their rhetorical origins is all very well but ultimately not a very constructive or fulfilling pastime. In this spirit, and to go some way to addressing the 'so what?' question that all academic researchers inevitably face, the final chapter of the book now offers some pointers for future practice, policymaking and research in the area of adult education and technology.

Chapter 11

Recommendations for future policy, practice and research

Introduction

We approach this final chapter from Bent Flyvbjerg's (2001) assertion that it is not enough for social researchers to investigate where society is going, who gains and who loses, and by what mechanisms of power. A critical approach, he argues, should also ask 'is it desirable?' and 'what should be done?'. In short, social research has a moral obligation to contribute to the general flow of public debate on our collective future. In this spirit our final chapter goes on to consider what aspects of adult learning in the digital age could, and should, be changed. It also discusses how any changes may be engineered. Our research has brought up many questions – for example, what aspects of adult education policymaking and practice should be altered or introduced anew and which aspects should be discarded or retained? Which government interventions should be allowed to continue as uncontrolled 'experiments' in the knowledge that they may well fail (Trow 1999)? To what extent will adult learning with ICTs simply continue to take place on an informal basis beyond the direct influence of the government or education community? We have tried to fashion responses to these issues which are reasoned and warranted, as adult education has a history of being hampered by rather too much change based on too little evidence. Since researchers, practitioners and policymakers inhabit relatively different worlds, it is not easy to pass on conclusions from research evidence (St Clair 2004), but based on the body of the empirical evidence that we have presented throughout this book, we feel in a better position than most to offer some suggestions.

Following Flyvbjerg's lead we first need to establish why ICT-based adult education is considered desirable before asking what should be done. Allied to the question of 'is it desirable?' is the question of 'does it matter?'. In terms of ICT-based adult learning in contemporary society our answer to the latter question would be a tentative 'no'. Engaging with ICT-based learning does not seem to profoundly influence adults' (immediate) lives in ways which are not achievable through other means, most especially via learning not involving ICT. In fact, failure to engage with either ICTs or education does not seem to hamper successful living in contemporary society. Although some radical commentators may like to portray a world where 'being disconnected is death' (Rifkin 2000: 187), this was not reflected in our research sample. Much of our data lead us to challenge the assumption made by some commentators that non-use of ICTs is a pressing twenty-first century source of 'marginality' and exclusion from 'substantial opportunities' (e.g. Castells 2001; Mossberger et al. 2003; Lenhart et al. 2003). For many of our respondents, computers were not something which directly impacted on their lives or that they felt compelled to engage with directly and actively. In short, the perceived 'desirability' for all citizens to use ICTs in order to survive and thrive in the current information age was not substantiated by the many people in our study who were surviving (and often thriving) without them. A similar conclusion can be drawn in terms of the perceived imperative for all citizens to engage in formal lifelong education. Direct engagement with formal education is still not an *essential* aspect of everyday life in contemporary society. As we have shown, none of the three main arguments for lifelong learning can be sustained in face of the experience of our respondents. It is not clear that jobs are becoming shorter, more flexible in contract or more multi-skilled than 50 years ago, for example. There are no obvious financial lifelong returns from learning, sometimes the reverse. And policies for lifelong learning, as with most studies of lifelong learning, appear to exclude the very social groups that they want to include.

We would advise, therefore, any parties concerned to increase engagement in ICT-based adult learning to first set about addressing the many non-technological issues which underpin non-engagement, such as poverty, housing, quality of employment and the family reproduction of inequalities. It is these factors, rather than any technology-driven issues which lie at the heart of the contemporary learning society. As Strover (2003: 275) concludes, academic presumptions about the significance of ICTs in everyday life lose credence 'in the face of no substantial change to our existing social structure attributable to computerisation'.

Yet although not essential to the current lives of many adults, we would stop short of arguing that ICTs and ICT-based learning should cease to be significant areas of interest for policymakers and educationalists. There is a danger of 'throwing the baby out with the bathwater' as reflected in the Chief of the US Federal Communication Commission's assertion that the digital divide was no more significant than the 'Mercedes divide' – i.e. a difference in terms of luxury goods and services which should not be seen as essential to people's well-being or survival (Powell 2001). Given their obvious educational potential we should not dismiss ICTs wholesale. Rather, we should be more circumspect in the importance that is attached to them.

Whilst we would concur that many in the population are living their everyday lives quite successfully without using computers *or* travelling in Mercedes cars, there remains a key question of when this non-engagement is the result of an empowered choice and when it may not be. In other words, when is an individual's non-engagement with ICTs and ICT-based learning the result of a fully informed and aware choice and not a result of structure and circumstance? Whereas we are all capable of making choices, it is a great deal easier for some people than others to make particular choices. We have seen throughout the empirical chapters of this book how some people are able to draw upon a wider range of economic, intellectual and social resources in making decisions to engage either with learning or technology. We would contend that there is a role for policymakers and educationalists to continue to nurture ICT-based adult learning because, unlike Mercedes driving, people have the right not to be excluded from access, even if they do not choose to take it up. Ensuring an equality of opportunity remains an important concern, so that the digital divide truly becomes a matter of digital choice.

That said, it is essential that this nurturing is carried out in more reasoned, nuanced, subtle and modest ways than at present. As we shall now go on to discuss, there is a need for those involved in adult learning in the digital age to (re)orientate their motivations and expectations along social and educational lines rather than with only deferred economic or political gratification in mind. There is also a need to 'go with the flow' of adults' lives and current patterns of learning rather than attempting to force radical new patterns of engagement. Developing a socially-grounded, realistic approach to adults' technology use should take place at all levels – from policymakers to practitioners, the IT industry to the research community. Our recommendations for each of these groups are discussed in the following sections.

Suggestions for government and policymakers

- Recognise that ICT is not a universal solution to adult non-participation.
- Do not privilege online learning over 'offline' learning – it is simply intended to overcome some of the barriers to the latter.
- Non-threatening sites for ICT-use, such as rural telecottages, should be strengthened rather than educational institutions as such.
- Community-driven, bottom-up initiatives – however temporary – should be respected, and not driven into certification routes.
- Allow technology to be used genuinely, as it would be in a private home, rather than for forced activities such as formal learning.
- Strengthen community resources so that technologies can be used at home rather than in public sites – thus equalising the quality of access.
- Encourage the creation of locally relevant and useful content.
- Recognise that all new technologies, such as digital TV, can create as many obstacles to learning as opportunities.
- Move towards a model of lifelong learning and ICT-use based on choice rather than deficit.

At the moment, the imperative underlying ICT-based learning and ICT-use exists largely in the eyes of the government rather than being based on tangible general public demand. As Caulkin (2004: 9) observes sceptically 'it is an article of faith that e-government is a good thing', rather than being driven by tangible evidence of effectiveness. This could be seen to confirm the neo-liberal suspicion that government has no right (or indeed need) to intervene in the supply and take-up of ICTs and, by extension, ICT-based learning. As we discussed in Chapter 1, many technologists believe strongly that as technologies converge and the cost of ownership falls then there will be an inevitable 'diffusion' effect where all but ideological 'refusniks' will have access to new ICTs as they approach 'full saturation' in domestic and work settings. Neo-liberal commentators would extend this argument to assert that the market is the most effective mechanism through which to distribute technologies. Indeed, as Navarra and Cornford (2003) observe, ICT reform in Western governments is usually implicitly grounded in a neo-liberal ideology of a reduction of the state. At most, the state should act only as a facilitator to ensure that markets flourish whilst the private sector is best placed to determine the scope and nature of the provision (see Moore 1998). The inability of governments to influence the development of ICT within their borders is compounded in the eyes of some commentators by the pressures of the global economy. As Rifkin (2000: 223) sees it, 'the deregulation and commercialisation of the world's telecommunications and broadcasting systems is stripping nation-states of their ability to oversee and control communications within their borders'. Thus, as Webster (2001: 268) reasons, governments which 'try to implement measures that militate against [technological] polarisation by redistributing resources to the most underachieving, risk losing the confidence of the world's markets which frown upon policies which introduce 'distortions'.

However, this view of social intervention and policymaking may be short-sighted. As John Hudson (2003: 275) argues, such claims 'ignore the depths of poverty ... and offer a grossly oversimplified view of the technology'. By definition, markets are less effective at distributing ICTs and ICT-based services to less 'profitable' individuals as well as shying away from products and services which are less commercially viable than others. This approach also ignores the remorseless game of catch-up being played as the power of technology advances and stratifies the quality of access across social groups. At present, the ineffective implementation of the information age and learning society policy agendas may not lie in the inappropriateness of government intervention *per se* but that governments are intervening in inappropriate and ineffective ways.

We would argue that governments should adopt more aggressive but also more realistic roles in encouraging citizen engagement with ICT and learning. This would involve focusing less on establishing a rhetorical universal access for all but instead aiming to enable effective access for all those who want it – when and where appropriate. This shifting of government attention from the high-profile, electorally attractive issue of universal service will be difficult, but is crucial if real progress is to be made. Although universal service is symbolically powerful it should not be confused with the real issues of the information age which are not party to uniform solutions (Sawhney 2000). On the basis of our research, effective enabling of the use of ICTs such as computers and the internet would seem to be better focused on supporting a genuine 'fit' with patterns of individuals' everyday lives – concentrating on increasing the relevance of ICT and shifting the universal service debate from issues of supply to issues of demand (Greenstein 1999). Crucially, 'demand' should be seen in terms of the genuine demands of individual adults rather than the usual policy conceptualisation of demand in terms of employers' demands for skills or governmental demand for economic competitiveness (Kingston 2004b).

As Rob Shields (2004) has observed, within the knowledge economy model the provision of all services – be they civic, social or educational – tends to be approached by governments as collective concerns with collective solutions. However, our research would suggest that policies should be more individually focused and less 'one-size-fits-all'. In short, governments need to extend choice on an individual basis to everyone regarding whether they participate in education and whether they use ICT to do so. Despite the occasional sudden impact of a charismatic mentor, people's decision to engage or not is generally based upon what an individual knows, how much time they have got, convenience, and how their tastes and preferences have developed early on in life. Everyone should have a chance to make an empowered choice to use ICT and to learn – but the overriding concern of policymakers has to be one of facilitating the individual's opportunity to choose rather than coercing mass engagement if they are to be successful in increasing and widening participation.

Unfortunately the current wave of policymaking in this area continues to pursue an avowedly collectivist path based around the inherent 'power' of ICT to prompt mass change, despite signs of the shortcomings of such an approach. High profile initiatives such as the University for Industry and learndirect are now facing an uncertain future as doubts over their effectiveness are raised by commentators other than us (Hook 2004b). Other authors are also beginning to question whether the education community is facing some of the worst facets of the 'dot com-foolery' that blighted e-commerce during the late 1990s – with a catastrophic 'boom and bust' affecting learners and investors alike (see, for example, the high profile demise of the UK 'e-University' project in 2004).

We can offer a number of tangible suggestions which may go some way to regaining this lost ground – first, with regard to the state-sponsored provision of ICTs for individuals otherwise without. The limited attraction of current sites such as schools, colleges and museums for facilitating ICT access suggests that different sites should be considered which may overcome some of the institutional barriers still in evidence. Previous research has highlighted the success of situating ICT access in 'non-threatening community-managed public place[s]' (Day and Harris 1997: 16) or, as Murdock (2004b) terms them, 'safe spaces'. Although the UK government has been keen to publicise the setting of centres in pubs, shopping centres and football stadia, in reality these locations represent a small proportion of existing provision (with *no* such provision in any of our twelve research areas). As Strover *et al.* (2004: 482) put it, the positioning of public ICT sites in 'convenient' sites like schools and libraries looks set to continue, if only 'for pragmatic and ... sometimes self-serving reasons ... [despite] the statistical and lived realities of the potential users'.

Yet it may well be that the currently tokenistic sites like pubs and football grounds stand more chance of truly widening public access to ICT and could be extended. Furthermore, some of the more successful public access sites in our research areas had been established 'below the radar' of official initiatives like UK Online, such as the less glossy and more idiosyncratic community-led and voluntary-run rural telecottages. Whilst the success of such centres would undoubtedly be compromised by a wholesale take-over and 'rebranding' by the government or UK Online they could be bolstered by financial support from central government. As Dondi (2003) argues, there should be genuine support and valorisation of 'bottom-up' community involvement by governments. Community-driven education should be given equal rights, funding and recognition.

Second, state-run public provision should be shaped around people's immediate needs for wanting to use the technology. At the moment, public provision of ICT is centred around the goals of gaining employment, lifelong learning and computer skills. As Hand (2005, in press) observes of library provision, 'at the institutional level, the internet-in-the-library is explicitly framed and understood in 'learning' terms: learning in new ways, with new means, for a series of specific ends ... promoting a specific kind of self-governing learner'. Whereas promoting ICT for 'learning' purposes may be attractive from a government and institutional level it results in a social positioning which, as we have seen, is often unattractive to some potential users.

We know that where people are given unfettered and undirected access to computers in 'genuine' community sites their engagement is often very different to what official bodies would deem desirable. Cunningham and Stover (2005) illustrate this point when reporting on a project in rural Texas where public terminals were appropriated by a number of different groups of previous non-users. Some of the heaviest patterns of usage included local forestry workers checking the daily weather forecasts and a range of people using the internet to view pornography. If the primary aim of this US project was to widen the use of the internet within the community then it should be deemed successful. Had this 'free' access been restricted to what was deemed acceptable or 'worthy' then it would undoubtedly have been less successful. To play devil's advocate for a moment, why should people's engagement with public ICT provision be any different from that of those who are able to access the internet at home or in work, where using the internet for checking the weather forecast, playing interactive games, using chatrooms, or accessing pornography is commonplace? If, as Dutton (2004) argues, the internet is 'an experience technology' that becomes gradually more established in people's lives the more that they use it, then encouraging any sort of use amongst current non-users would seem sensible. A fully participative information society will not arise from mass use of the internet solely for education or civic engagement.

Returning to the issue of hardware provision, a third suggestion is to refocus state efforts away from community *sites* towards developing systems of community *resources.* There is a need, perhaps, to rethink state efforts to facilitate use of ICT by adults and, in particular, to explore the possibilities of re-appropriating community ICT provision into private, domestic settings rather than municipal, public sites. A wider range of sites could be considered where people can access and make use of ICT. For example, systems of community resources could be developed which can then be loaned into people's houses – thus building upon adults' apparent willingness to use ICT in their own homes and the homes of their extended families. This approach makes increasing sense as ICTs such as handheld computers and personal desktop assistants become more mobile and less reliant on 'fixed' sites and physical connection to power and telecommunications networks. Indeed, the notion of loaning community resources has been tentatively explored through pilot initiatives in the UK such as 'Computers within Reach' which involved the placement of refurbished computers in 100,000 socially excluded households, and the 'Wired Up Communities' initiative which involved the connecting of homes in seven different 'disadvantaged communities' in England to the internet. Despite initial evaluations of these innovative initiatives highlighting the difficulties in encouraging widespread *use* of ICTs once having provided access (Policy Research Institute 2002; Devins *et al.* 2003), we would whole-heartedly recommend the wider adoption of this particular approach.

Of course, simply providing equipment solves very little in itself. There are many examples in education where the supply of computers does nothing to solve pre-existing problems, even in countries with the most computers in educational institutions (such as Sweden, see Debande 2004). As our earlier Texan example illustrated, there is also a need for state provision to engage realistically with what people can do with ICT once they have been provided access to it. At the heart of getting citizens engaged and involved with the digital age is making ICTs and ICT-based services useful and relevant to people. Yet as we have seen from our research, ensuring that online services are genuinely useful and relevant to people's everyday lives is an issue which remains unresolved throughout the UK government's current policy drive. From this point of view it would seem appropriate for the government and other interested parties to begin to consider alternative means of 'reshaping' ICT to fit better with the lives of those adults who are not using ICT – rather than the other way around. The point has been well made recently that many government websites purporting to offer citizens ready access to state services such as pensions, social security, television licensing and the like are underused due to their lack of substance and utility to the individuals they are designed for (Hedra 2002; Public Accounts Committee 2002). Indeed, the lack of locally-relevant and locally-useful content remains a consistent criticism of the current forms of 'e-government' content in the UK (Olsen 2004).

There is an opportunity here to encourage locally-based creation and development of content and therefore avoid the situation where 'people and groups within a settlement or community are forced into some imposed and standardised 'top-down' model which neglects the huge diversity of communicational cultures between them' (Graham 2002: 52). In terms of 'educational' content the scope for encouraging community creation of content is endless, from local history and arts activities to genealogy and gardening. Whilst community creation of content is obviously a 'bottom-up' piecemeal process, government can lead by example by being good producers and users of ICT themselves – creating and maintaining genuinely useful and relevant websites and online services and not being seen to merely pay lip-service to ICT. If our own experience of trying to find out the location of the learndirect sites in our twelve research areas was anything to go by, this remains a considerable challenge.

Perhaps the biggest shift required in policy thinking is the need to recognise that the creation of inclusive information and learning societies is not an issue of funding and technology but hinges around wider problems of exclusion from society which the public provision of ICT can be expected to alter little. There is an implicit technological determinism in the assumption that public access to ICTs will impact on people's social inclusion, educational levels and employment. It is, of course, more likely to be the case that people's socio-economic status impacts on their opportunities and need to use ICTs. As it currently stands, state ICT policies are 'failing' to achieve their wider inclusive aims as they are predicated on a fundamental political misunderstanding about the nature of the digital divide. As Ellis (2001: 9) argues, 'the real divide is, and always has been, a socio-economic one'. Put simply, what excluded individuals need first may not be access to ICT but alleviation from the basic causes of poverty and exclusion. As Starr observes:

> by focussing on a symptom of poverty – lack of access to information technology – the champions of online community evade the obligation to do something substantial about poverty. And by appointing themselves as saviours, they end up interfering in people's informal, everyday relationships.
> (Starr 2002)

Above all, effective digital-age policymaking requires a recognition that the underlying causes of exclusion from the information society are the same non-technological issues which underlie exclusion from society in general. Otherwise, state ICT provision will continue to have the modest impacts on the adult population which this book has outlined.

Unfortunately we make these suggestions for a realistic and pragmatic approach to policymaking more in hope than expectation. Most of these ideas are not new, but neither are they currently being pursued with any vigour. The current policy context shows few signs of deviating from its established top-down, coercive approach which we see little hope of working. For example, although adults and ICT continue to be at the forefront of UK government thinking, a 'deficit' view of non-use persists where universal use remains a case of overcoming technical deficits and finding the right way of going about it. The current 'killer application' in this respect is seen to be digital television. In some areas the government has begun to divert their attention away from computer-based content to digital television via initiatives such as NHS Direct Digital TV. Yet although the penetration of digital television will soon well exceed that of computers, the key issues of (ir)relevance and interest remain.

Indeed, one could argue that digital TV will further marginalise activities such as learning, as the mass-market continues to demand what Schofield-Clark *et al.* (2004: 544) bemoan as 'enhanced leisure-orientated consumption' as opposed to 'desired educational and civic uses' of technologies. Faced with 50 channels of football and films it could be all too tempting for someone to postpone tuning in to the online learning channel for another evening. And the increasing specialisation of channels is changing our TV-viewing behaviour, making it less likely that we stumble across and are entranced by an 'educational' programme than when faced with the traditional mixed fare of terrestrial channels. The fact that some adults do not see a need either to use ICT or engage in learning is a long way from being recognised in policy terms. For example, the current £18 million 'Cybrarian' government project continues the trend of seeing those who do not use the internet purely as those who either 'lack the confidence and skills, or have mental and physical disabilities' (Lamb 2004: 9). One exception to this was the Cabinet Office's 'Digitally United Kingdom' report which acknowledged that:

> supply-side solutions alone will not successfully shift the entire UK population to digital engagement – no matter how hard we try. Very simply, some people will make a rational decision to remain digitally unengaged, while others may not be able to be engaged because of their own particular circumstances: such people may prefer instead to use traditional channels of communication.
>
> (Cabinet Office's 2004: 39)

But even this perceptive report retained a faith in coercive approaches to boosting levels of use through 'appropriate' marketing and development.

Similarly, even the more 'cutting-edge' academic thinking in this area tends to be predicated upon the inherent desirability of ICT-use and learning – leading us to conclude that it is not that policymakers are lacking ideas but lacking a fundamental direction and basis to their efforts. For example, Mossberger *et al.* (2003) argued recently for the establishment of educational vouchers or an 'educational technology subsidy' which provides money for purchasing computer equipment linked to the stipulation that the recipient participates in some form of further education. The logic for this is 'connecting the voucher to educational advancement would also emphasise the public good potential of information technology, addressing the educational issues that surfaced in both the economic opportunity divide and the democratic divide' (p. 136). However, this again assumes a willingness to be coerced into learning which is unlikely to prove attractive to those who are currently non-learners. Another recently mooted policy option is the idea of 'zero stop shops' (Margetts and Dunleavy 2002) which aim to seek out those not accessing help for which they are entitled. This is in contrast to the current 'one-stop shop' model of public access which merely makes opportunities available from one point. Again there is little to suggest that this approach will not encounter similar barriers of irrelevance and non-interest as before – rooted as it is in a model of coercion and compulsion.

Suggestions for the IT industry

- Policymakers and manufacturers must strengthen their links in recognition of their inter-reliance.
- Educational software and applications should be wider in scope than simply office functions, and bundled into hardware packages as a matter of course.
- We need cheaper platforms, rather than necessarily always more powerful ones.
- We could import from other developed countries, such as the US, that culture of philanthropic giving on the part of IT companies to overcome the digital divide.

Although important, government policies alone do not constitute 'educational computing'. However keen governments may be for adults to make use of computers, they are incapable of supplying and providing the hardware and software themselves. Technically and financially the mass production of computers is beyond the means of the state and, in all but the most dirigiste countries, is a multi-billion pound business left to the private sector. For these supply-side reasons alone IT firms and the 'IT industry' are also integral elements of adults' use of ICT. Indeed, adult learning in the digital age is a good example of Kenway et al.'s (1994) 'markets/education/technology' triad. Although initially sponsored by the state, ICT-based adult learning is largely reliant on commercial interests, public and private expertise and is overtly informed by both state *and* market values. The centrality of IT firms and private interests to the educational use of ICT is obvious but it is more often than not ignored by academic observers. It is assumed that IT firms are a neutral or even benign presence in adults' use of ICT – quietly providing hardware and software and then slipping away. Yet the centrality of private concerns in the very public arena of educational technology merits sustained attention – especially regarding how adults' use of ICTs can be shaped by business and commercial interests.

The IT industry was implicit in much of our study: shaping respondents' expectations of what ICTs could and could not be used for; (mis)selling machines; operating helplines; and bundling software in with computers. Put simply, ICT-based adult learning *needs* IT firms to be successful. The question we have to ask is how much do IT firms need to cater for the needs of adult learning? Moreover, is the involvement of private interests in adult learning capable of positively shaping it? In theory there are many ways in which the IT industry could alter their practices and thereby encourage widespread adult use of ICT and ICT-based learning. IT firms have a clear role in shaping adults' initial uses of computers from the marketing through to the bundling of free software at the initial point of purchase. At best, currently, an encyclopaedia such as Encarta may be included at the point of purchase of a new computer but there is scope here for a little more creative thought on the part of IT firms and adult educators. On a prosaic level the large IT companies could pay more attention to the content that is supplied alongside their hardware – developing useful software for all adults which can then be supplied with machines – not just endless variations on office applications.

Again, following on from the policy recommendation of governments providing shared public access to ICT through domestic networks, the notion of IT firms also supplying subsidised or even free computers to those social groups currently without computers is an attractive idea. Whereas redistribution and recycling schemes are currently in operation in most countries they are generally localised and relatively small-scale. With the sustained involvement of national and multi-national firms, the widespread loaning or even donation of computers, internet connections and software to individuals, households and other informal settings could take place. A PC could, thus, become the low-cost vehicle, like a TV, for a range of services that are themselves the basis for profit. Firms face any suggestions of 'social responsibility' with understandable ambiguity (Bennett 1995) and there is little short-term incentive for IT firms to engage in any of these activities. For this reason alone links between policymakers and industry must continue to be fostered and maintained, as Schofield-Clark et al. reason:

> Policymakers and others concerned with the digital divide must confront the problem of how to encourage positive civic and educational uses without giving unfettered freedoms to the corporations who stand not only to [financially] benefit ... but are absolved of any social responsibility.
>
> (Schofield-Clark et al. 2004: 544)

Yet there is no reason why such activity could not be attractive to firms given their vested interest in encouraging computer use. As Dyson (1998: 122) observes, 'one of the problems with [educational funding] is that governments and philanthropists often feel good simply by giving it away; investors feel good when the money they invest actually produces something'. In this respect firms are in a position to contribute to the establishment of a wider market for ICT which they ultimately stand to benefit from. Indeed, the IT industry is not adverse to such philanthropy but have thus far preferred to concentrate on schools and developing countries. In the US the digital divide is a growing area for philanthropic activity by multi-national companies and foundations such as the Bill and Melinda Gates Foundation, IBM, Cisco and AOL/Time Warner. We would suggest that this activity could easily be extended to the provision for adults in the UK and other developed countries.

Suggestions for practitioners

- Refocus formal educational provision away from learning *about* ICT and towards learning with ICT.
- We should learn from formal systems analysis not to automate our errors – ICT in education does not make up for poor practice.
- Move the focus of our efforts from a predominantly schools, FE and HE emphasis to *lifelong* learning for all.
- Let go of an obsession with teaching, especially in institutions, in favour of a recognition of learning.
- Giving informal learning a higher status is equivalent to widening participation at a stroke.
- The best technological champions in real communities could be the new learners.

Despite the apparent importance of lifelong participation to the individual (Schuller *et al.* 2004; Feinstein and Hammond 2004), we still know relatively little about what works in this regard (Macleod 2003). Although many of our findings point towards the relative unimportance of formal adult education provision within current patterns of ICT-based learning, there are a number of ways in which providers and practitioners can better engage with adult learning in the digital age and give it a crucial element of 'legitimacy' with stakeholders *and* end-users (Navarra and Cornford 2003).

First and foremost is the need to reshape formal educational provision away from learning *about* ICT and towards learning with ICT. Providing extra places and courses, and making participation easier and cheaper are certainly necessary precursors to improving rates of participation among adults, but there is a danger that adult education is being seduced by ICT and 'e-learning', whilst using the technology to replicate existing bad practice. We have seen from our data that people learn best through using ICT for something else and with others. This suggests, therefore, that fewer ICT courses and more courses with ICT would be more attractive to more people as well as leading to the development of 'really useful' ICT skills. Indeed, there is growing evidence that computer skills education is of little long-term benefit, even in terms of the often-argued 'employability' gains. Recent detailed economic analysis of UK labour-force data has revealed the low payback of learning computer skills for lower skilled workers – especially when compared to the advantages gained from improving basic literacy and numeracy skills. As Borghans and Weel concluded:

> neither the increase in computer use nor the fact that particularly higher skilled workers use a computer provides evidence that computer skills are valuable … Results show that while the ability to [manually] write documents and to [manually] carry out mathematical analyses yields significant labour-market returns, the ability to effectively use a computer has no substantial impact on wages.
>
> (Borghans and Weel 2004: 85)

There is also a need not to focus purely on specific target groups of the potentially economically active in the name of social inclusion. It has been argued that the UK is presently experiencing a decline in learning stemming from the government's focus on specific groups. As Alan Tuckett argues, 'the relentless focus of funders on achievement targets is narrowing the curriculum offered to adults, as expansion of provision is bought at the expense of their elders' (cited in Kingston 2004a: 15). This is evident in the avowedly workplace-focused provision on offer through learndirect and many other e-learning providers. The findings of this study also suggest that the determinants of early and later participation are significantly different, and that simply increasing front-loaded provision (increasing further and higher education, for example) is unlikely to 'cash out' into increasing lifelong learning trajectories. This raises the crucial policy issue of where scarce resources for education and training should be directed, especially given the focus up until now on 'front-loading' investment into initial schooling. Whilst the evidence should be treated with caution, it does indicate that shifting this balance in favour of policies addressed to the determinants of later partici-pation would be more efficient and cost-effective.

Finally, formal education should be seen as a good place to harness the potential of interested learners to diffuse their learning and ICT-use through their own informal networks. We would argue from our data that one of the most effective ways in which practitioners can widen adult participation in education is by encouraging and supporting informal learning outside formal settings. This would involve an altruistic turn within the sector to facilitate learning which does not necessarily lead to any further learning or engagement in formal education but merely allows people to learn what they want. Although there will be little or no short-term financial reward in doing so, the deferred benefit of stimulating adults into learning (however informal) is obvious. With regard to encouraging computer use, formal education providers should also be looking to create 'technological champions' of the adults who do choose to come through their doors and take an IT course. These individuals are obviously motivated but can also play a role in 'cascading' IT skills and use through their informal networks of family, friends and colleagues (see Blythe and Monk 2005). Given that we know that most people learn to use computers through informal sources, then encouraging and supporting adult learners to act in this way could be a key means to widening engagement with ICT. At present, however, the talents of these potential champions are too often used in the very limited setting of centres as tutors for each new cohort.

Suggestions for researchers

- Research on both ICT and adult learning needs strengthening.
- We suggest a greater variety of methods, including longitudinal, experimental and mixed methods approaches at various levels of analysis.
- Research needs to move away from providers, and routinely involve com-parison groups such as non-users and non-learners.

- A variety of topics would repay further work, including self-directed learning, the world of non-users, and the role of place in patterns of use and non-use.

Finally, no research project would be complete without recommendations and suggestions for other researchers. Although it is a cliché for academics to conclude that more research work is needed, we would contend that in the case of adult learning in the digital age more research is genuinely required. If nothing else this book has highlighted the need for sustained attention to be given to the area of adult learning and ICT – however unfashionable and marginal adult learning may be for mainstream social science researchers. There is a pressing need for the development of a body of high quality, rigorous research – using a range of methods and designs.

Barry Wellman argues that social studies of new technologies have progressed from an initial stage of 'punditry riding rampant' through to careful and systematic documentation of users and uses. However, he urges researchers to move beyond this 'low hanging fruit' and go on to pursue a 'real analysis' and 'more focused, theoretically driven projects' (Wellman 2004: 127). Engineering a longitudinal element, as we have here, into future research appears crucial. We need to move studies away from a focusing only on the providers and on the participants in lifelong learning. And, as Hargittai (2004: 141) argues, 'we need to expand studies … to units of analysis other than the individual'. These units of analysis can take the form of levels of study which focus on the contexts of use highlighted throughout this book. For example, to what extent do households, communities, families and workplaces have access to and use ICT-based learning? There are a range of different perspectives which researchers should consider adopting. For example, we need more research on the phenomenological experience of learning with ICTs. As Murdock (2004b: 23) contends, 'analysis of the repercussions of structural forces of change needs to be grounded in detailed ethnographic work on everyday action'. Conversely, and perhaps most importantly, adult education is an area devoid of experimental approaches meaning that theorists never have to put their cherished ideas at risk in a randomised trial or similar.

Many areas have emerged which are worthy of future investigation. First, the role of informal learning was a recurrent theme throughout our data and is deserving of more sustained and subtle investigation – in particular the synergy between informal learning at home and in the workplace. Second, non-users of computers and the internet emerged from our project as a large but currently under-researched group of the general population. More ethnographic work on non-users of ICTs is needed which will explore the agency and structure underlying individuals' sustained non-engagement with new technologies in day-to-day life. Finally, the role of place and geography was highlighted repeatedly in our survey and case-study data in terms of educational participation and engagement with ICTs. Whilst our current data have gone some way in exploring this it is clear that more focused research is required, by ourselves and others, to fully explore the influence of place on ICT-use and learning.

Conclusions

The following points recur throughout the previous set of suggestions to policymakers, IT industry, adult educationalists and researchers. In short, at all levels we are calling for:

- a readjustment of expectations when approaching adult learning in the digital age;
- recognising that there is no homogeneous 'learner' or 'non-learner' for whom a generalisable model of best practice can be developed;
- developing any action or intervention from a 'bottom-up' perspective and centred around fitting with the individual's everyday life;
- giving people, if possible, a genuine need to learn rather than assuming the impetus already exists;
- giving people, if possible, a genuine need to use ICT;
- ensuring that people have effective access to ICT;
- not attempting to coerce people into using ICT but supporting those who do;
- drawing on people's propensity to use informal means of engaging with technology and education.

Above all we are calling for a little less enthusiasm and a little more realism when it comes to approaching adult learning in the digital age. Whilst it is perhaps understandable for governments and technologists to overestimate the gains of applying ICTs to the delivery of public services whilst downplaying the conflicts and complexities involved, adopting a naïve optimism will be of little benefit to achieving the *effective* integration of ICTs into adult education. There is no longer a need for technologists and politicians to 'sell' new technologies to a disbelieving educational community. What is needed is a realistic approach to technology and adult education which avoids the damaging but well-worn cycle of 'hype', 'hope' and eventual disappointment as yet another educational technology fails to live up to initial expectations.

If these changes are made then we may well achieve a le@rning society, although it may not be one like the one that we are currently expecting or imagining. The likely le@rning society of the near-future will not be easily measurable and bench-markable; it will not involve everyone at all times during their life and it will not overshadow the very important basic and often prosaic issues which need to be addressed in order to right society's wrongs (poverty, illiteracy and so on). It will not always (or even often) be transformative. But it will see ICT helping those who want or feel that they need to engage in learning. That is probably as good as we can hope for.

References

Abroms, L. and Goldschieder, F. (2002) 'More work for mother: how spouses, cohabiting partners and relatives affect the hours mothers work', *Journal of Family and Economic Issues*, 23, 2: 147–66.

Alger, J. (2001) 'Legal issues in the e-learning business', paper presented at the EDUCAUSE 2001 annual conference, Indianapolis, Indiana, 28–31 October.

Anderson, A. and Nicol, M. (2000) 'Computer-assisted vs. teacher-directed teaching of numeracy in adults', *Journal of Computer Assisted Learning*, 16, 3: 184–92.

Aune, M. (1996) 'The computer in everyday life: patterns of domestication of a new technology', in M. Lie and K. Sorensen (eds) *'Making Technology our Own?'*, Oslo: Scandinavian University Press.

Aveling, N. (2002) 'Having it all and the discourse of equal opportunity: reflections on choices and changing perceptions', *Gender and Education*, 14, 3: 265–80.

Bakardjieva, M. (2001) 'Becoming a domestic internet user', *Proceedings of the Third International Conference on Uses and Services in Telecommunications*, Paris: France Telecom.

Ball, S. (1999) 'Labour, learning and the economy: a policy sociology perspective', *Cambridge Journal of Education*, 29, 2: 195–206.

Balnaves, M. and Caputi, P. (1997) 'Technological wealth and the evaluation of information poverty', *Media International Australia*, 83: 92–102.

Banks, M., Bates, I., Bynner, J., Breakwell, G., Roberts, K., Emler, N., Jamieson, L. and Roberts, K. (1992) *Careers and Identities*, Milton Keynes: Open University Press.

Becker, G. (1975) *Human Capital: A Theoretical and Empirical Analysis*, Chicago: University of Chicago Press.

Beinart, S. and Smith, P. (1998) *National Adult Learning Survey 1997*, Sudbury: Department for Education and Employment.

Bell, D. (1973) *'The Coming of Post-industrial Society*, New York: Basic Books.

Bennett, R. (1995) 'School-business links: clarifying objectives and processes', *Policy Studies*, 16, 1: 23–48.

Benton Foundation (1996) *The Learning Connection: Will the Information Highway Transform Schools and Prepare Students for the Twenty-first Century?*, Washington, DC: Benton Foundation.

Berg, P., Kalleberg, A. and Appelbaum, E. (2003) 'Balancing work and family: the role of high commitment environments', *Industrial Relations*, 42, 2: 168–88.

Berge, Z. and Muilenburg, L. (2004) 'Results of pilot survey on student barriers to e-learning', paper presented to the Tele-learning 2004 Conference, San Diego, CA, 21–24 February.

Berghman, J. (1995) 'Social exclusion in Europe: policy context and analytical framework', in G. Room (ed.) *Beyond the Threshold: The Measurement and Aanalysis of Social Exclusion*, Bristol: Policy Press.

Berman, Y. and Phillips, D. (2001) 'Information and social quality', paper presented to the Fifth European Sociological Association Conference, University of Helsinki, Finland, 28 August–1 September.

Bijker, W., Hughes, T. and Pinch, T. (1987) *The Social Construction of Technological Systems*, Cambridge, MA: MIT Press.

Blanton, W.E., Moorman, G. and Trathen, W. (1998) 'Telecommunications and teacher education: a social constructivist review', *Review of Research in Education*, 23: 235–75.

Blaxter, L. and Tight, M. (1995) 'Life transitions and educational participation by adults', *International Journal of Lifelong Education*, 14, 3: 231–46.

Blythe, M. and Monk, A. (2005) 'Net neighbours: adapting HCI methods to cross the digital divide', *Interacting With Computers*, 17, 1: 35–56.

Bonfadelli, H. (2002) 'The internet and knowledge gaps: a theoretical and empirical investigation', *European Journal of Communication*, 17, 1: 65–84.

Borghans, L. and ter Weel, B. (2004) 'Are computer skills the new basic skills? The returns to computer, writing and math skills in Britain', *Labour Economics*, 11, 1: 85–98.

Borgida, E., Sullivan, J., Oxendine, A., Jackson, M., Riedel, E. and Gangl, A. (2002) 'Civic culture meets the digital divide: the role of community electronic networks', *Journal of Social Issues*, 58, 1: 125–42.

Bostock, S. (1998) 'Constructivism in mass higher education: a case study', *British Journal of Educational Technology*, 29, 3: 225–40.

Bromley, C. (2004) 'Exploring digital dynamics: findings from the British Social Attitudes survey', paper given to OfCOM consumer panel seminar, London, 29 November.

Bromley, H. (1997) 'The social chicken and the technological egg: educational computing and the technology/society divide', *Educational Theory*, 47, 1: 51–65.

Brown, A. (1992) 'Design experiments: theoretical and methodological challenges in creating complex interventions in classroom settings', *Journal of the Learning Sciences*, 2, 2: 141–78.

Brown, J. and Duguid, P. (1991) 'Organisational learning and communities of practice', *Organisation Science*, 2, 1: 40–57.

Brown, P. and Lauder, H. (1996) 'Education, globalization and economic development', *Journal of Education Policy*, 11: 1–25.

Bruland, K. (1995) 'Patterns of resistance to new technologies in Scandinavia: an historical perspective', in M. Bauer (ed.) *'Resistance to new Technology'*, Cambridge: Cambridge University Press.

Burke, C. (2003) 'Women, guilt and home computers', in J. Turow and A. Kavanaugh, (eds) *The Wired Homestead: An MIT Sourcebook on the Internet and the Family*, Cambridge, MA: MIT Press.

Burrows, R. and Bradshaw, J. (2001) 'Evidence-based policy and practice', *Environment and Planning A*, 33: 1345–8.

Burstall, E. (1996) 'Second chance stifled', *Times Educational Supplement*, 22 March: 27.

Butler, P. (1999) 'Just three years left', *t Magazine*, November: 19.

Cabinet Office (2000) 'Government to speed up introduction of online services', *Press Release CAS 140/00*, 30 March.

Cabinet Office (2004) *Enabling a Digitally United Kingdom*, London: The Stationery Office.

Calvert, S., Rideout, V., Woolard, J., Barr, R. and Strouse, G. (2005) 'Age, ethnicity, and socioeconomic patterns in early computer use – a national survey', *American Behavioral Scientist*, 48, 5: 590–607.

Campbell, C. (1992) 'The desire for the new', in R. Silverstone and E. Hirsch (eds) *Consuming Technologies*, London: Routledge.

Caron, A. and Caronia, L. (2001) 'Active users and active objects: the mutual construction of families and communication technologies', *Convergence*, 7, 3: 38–61.

Casino, M., Sklar, B., Nguyen, H., Just, M., Galzagorry, G. and Bakken, S. (2002) 'Implementing a web-based information resource at an inner-city community church', *Computers Informatics Nursing*, 20, 6: 244–50.

Cassidy, M. (2001) 'Cyberspace meets domestic space: personal computers, women's work, and the gendered territories of the family home', *Critical Studies In Media Communication* 18, 1: 44–65.

Castells, M. (1996) *The Information Age: Economy, Society and Culture. Volume I: The Rise of the Network Society*, Oxford: Blackwell.

Castells, M. (1999) 'An introduction to the information age', in H. Mackay and T. O'Sullivan (eds) *The Media Reader: Continuity and Transformation*, London: Sage.

Castells, M. (2001) *The Internet Galaxy: Reflections on the Internet, Business and Society*, Oxford: Oxford University Press.

Caulkin, S. (2004) 'E-binge which will cost us dear', *The Observer* (Business section), 15 August: 9.

Central Office of Information (1998) *Our Information Age: The Government's Vision*, London: The Stationery Office.

Chatman E. (1996) 'The impoverished life-world of outsiders', *Journal of the American Society for Information Science*, 47, 3: 193–206.

Chew, F. (1994) 'The relationship of information needs to issue relevance and media use', *Journalism Quarterly*, 71: 676–88.

Choi, J. (2002) 'Embedding digital television in an IT economy', *The Journal of International Communication*, 8, 2: 26–45.

Clement, A. and Shade, L. (2000) 'The access rainbow: conceptualising universal access to the information/communication infrastructure', in M. Gurstein (ed.) *Community Informatics*, Hershey, PA: Idea Publishing.

Coffield, F. (1994) *Research Specification for the ESRC Learning Society: Knowledge and Skills for Employment Programme*, Swindon: ESRC.

Coffield, F. (1997) *A National Strategy for Lifelong Learning*, Newcastle upon Tyne: Department of Education, University of Newcastle.

Coffield, F. (1999) 'Breaking the consensus: lifelong learning as social control', *British Educational Research Journal*, 25: 479–500.

Commonwealth Department of Education, Science and Training (DEST) (2001) *Lifelong Learning: Demand and Supply Issues: Some Questions for Research*, Canberra: Commonwealth Department of Education, Science and Training.

Connolly, M., Saunders, D. and Hodson, P. (2001) 'Can computer-based learning support adult learners?', *Journal of Further and Higher Education*, 25, 3: 325–35.

Cook, J. and Smith, M. (2002) *'Final Report for Study of UK Online Centres'*, London: London Metropolitan University, Learning Technology Research Institute.

Cornford, J. and Pollock, N. (2003) *Putting the University Online: Information, Technology and Organisational Change*, Buckingham: Open University Press.

Council of the European Union (2002) 'Council decision of 18 February 2002 on guidelines for member states' employment policies for the year 2002', (2002/177/EC) *Official Journal of the European Communities*, L60/60, 1 March.

Crace, J. (2004) 'Open for business', *The Guardian*, Education supplement, June 8 : 15.

Cranmer, S. (2002) 'Digital overflow: negotiating the demands of the work place using the internet at home', paper presented to the Internet 3.0: Net/Work/Theory conference, Maastricht, October.

Creswell, J. (2003) *Research Design: Qualitative, Quantitative and Mixed Methods Approaches*, Thousand Oaks, CA: Sage.

Crook, C. (1998) 'Children as computer users: the case of collaborative learning', *Computers and Education*, 30, 3/4: 237–47.

Cullen, J. (2000) *Informal Learning and Widening Participation*, Nottingham: Department for Education and Employment.

Cullen, J., Hadjivassiliou, K., Kelleher, J., Sommerlad, E. and Stern, E. (2002) *Review of Current Pedagogic Research and Practice in the Fields of Post-compulsory Education and Lifelong Learning*, Report submitted to the Economic and Social Research Institute, London: Tavistock Institute.

Cunningham, C. and Stover, S. (2005) 'The possibilities and limits of community networking', in P. Golding and G. Murdock (eds) *Unpacking Digital Dynamics: Participation, Control and Exclusion*, Cresskill, NJ: Hampton Press.

Curtain, R. (2002) *Online Delivery in the Vocational Education and Training Sector: Improving Cost Effectiveness*, Adelaide: National Centre for Vocational Education Research.

Daley, B., Watkins, K., Williams, S., Courtenay, B., Davis, M. and Dymock, D. (2001) 'Exploring learning in a technology-enhanced environment', *Educational Technology and Society*, 4, 3: 126–38.

Day, P. and Harris, K. (1997) *Down to Earth Vision: Community-based IT Initiatives and Social Inclusion*, London: IBM.

De Kerckhove, D. (1997) *Connected Intelligence: The Arrival of the Web Society*, London: Kogan Page.

Dearing, R. (1997) *Higher Education in the Learning Society: Report of the National Committee into Higher Education*, London: HMSO.

Debande, O. (2004) 'ICTs and the development of eLearning in Europe: the role of the public and private sectors', *European Journal of Education*, 39, 2: 191–203.

Dede, C. (1981) 'Educational and social implications', *Programmed Learning and Educational Technology*, November 1981; reprinted in T. Forester (ed.) *The Information Technology Revolution*, London: Basil Blackwell.

Denholm, J. and Macleod, D. (2003) *Prospects for Growth in Further Education*, Wellington: Learning and Skills Research Centre.

Department for Education and Employment (DfEE) (1996) *Skills and Enterprise Executive Issue 2/96*, Nottingham: Skills and Enterprise Network.

Department for Education and Employment (DfEE) (1998) *The Learning Age: A Renaissance for a New Britain*, London: The Stationery Office.

Department for Education and Employment (DfEE) (1999) *Learning to Succeed: A New Framework for Post-16 Learning*, London: The Stationery Office.

Department for Education and Skills (DfES) (2002a) *Get on with IT: The Post-16 e-Learning Strategy Task Force Report*, London: Department for Education and Skills.

Department for Education and Skills (DfES) (2002b) *Success for All: Reforming Further Education and Training*, London: Department for Education and Skills.

Devine, K. (2001) 'Bridging the digital divide', *Scientist*, 15, 1: 28.

Devins, D., Darlow, A. and Smith, V. (2002) 'Lifelong learning and digital exclusion: lessons from the evaluation of an ICT Learning Centre and an emerging research agenda', *Regional Studies*, 36, 8: 941–5.

Devins, D., Darlow, A., Burden, T. and Petrie, A. (2003) *Connecting Communities to the Internet: Evaluation of the Wired-up Communities Programme 2000–2002*, London: Department for Education and Skills.

Dhillon, J. (2004) 'An exploration of adult learners' perspectives of using learndirect centres as sites of learning', *Research in Post-Compulsory Education*, 9, 1: 147–58.

Dhunpath, R. (2000) 'Life history methodology: "narradigm" regained', *International Journal of Qualitative Studies in Education*, 13, 5: 543–51.

Dickinson, P. and Sciadas, G. (1999) 'Canadians connected', *Canadian Economic Observer*, 3: 1–22.

Dillon, P. (2004) 'Trajectories and tensions in the theory of information and communications technology in education', *British Journal of Educational Studies*, 52, 2: 138–50.

Dirkx, J.M. and Taylor, E. (2001) 'Computer technology and teaching adults: beliefs and perspectives of practising adult educators', paper presented at the National Adult Education Conference, Baltimore, MD, 18 October.

Doctor, R. (1994) 'Seeking equity in the national information infrastructure', *Internet Research: Electronic Networking Applications and Policy*, 4, 3: 9–22.

Dondi, C. (2003) 'ICT and adult and informal learning in the EU', paper presented to the OECD/NCAL International Roundtable, Philadelphia, 12–14 November.

Doring, A. (1999) 'Information overload', *Adults Learning*, 10, 10: 8–11.

Doubler, S., Harlen, W., Harlen, W., Paget, K. and Asbell-Clarke, J. (2003) 'When learners learn on-line, what does the facilitator do?', paper presented at the American Educational Research Association Annual Conference, Chicago, IL, 21–25 April.

Douglas, J. (1992) 'Freedom of education will solve our educational crisis', *The Freeman*, 42: 6.

Dutton, W. (2004) *Social Transformation for the Information Society*, Paris: UNESCO WSIS publication series.

Dyson, E. (1998) *Release 2.1: A Design for Living in the Digital Age*, London: Penguin.

Edwards, R. (1993) *Mature Women Students: Separating or Connecting Family and Education*, London: Taylor & Francis.

Edwards, R. and Usher, R. (1998) 'Lo(o)s(en)ing the boundaries: from "education" to "lifelong learning"', *Studies in Continuing Education*, 20, 1: 83–103.

Edwards-Johnson, A. (2000) 'Closing the digital divide', *Journal of Government Information*, 27, 6: 898–900.

Ellis, V. (2001) 'Enterprise or exploitation: can global business become a force for good?', *New Statesman*, 16 July: N1–16.

Erzberger, C. and Prein, G. (1997) 'Triangulation: validity and empirically-based hypothesis construction', *Quality and Quantity*, 31: 141–54.

E-Skills NTO (2001) *Telecommunications Strategic Plan for the UK, 2002–2005*, London: E-Skills National Training Organisation.

European Commission (2001) *Making a European Area of Lifelong Learning a Reality*, Brussels: DG Education and Culture, European Commission Communication.

Faber, J. and Scheper, W. (2003) 'Social scientific explanations?', *Quality and Quantity*, 37: 135–203.

Facer, K., Furlong, J., Furlong, R. and Sutherland, R. (2001) 'Home is where the hardware is: young people, the domestic environment and access to new technologies', in I. Hutchby and J. Moran-Ellis (eds) *Children, Technology and Culture*, London: Falmer Press.

Facer, K., Furlong, J., Furlong, R. and Sutherland, R. (2003) *Screenplay*, London: Routledge.

Faure, E., Herrera, F., Kaddoura, A., Lopes, H., Petrovsky, A., Rahnema, M. and Champion Ward, F. (1972) *Learning to Be: The World of Education Today and Tomorrow*, Paris: UNESCO.

Feinstein, L. and Hammond, C. (2004) 'The contribution of adult learning to health and social capital', *Oxford Review of Education*, 30, 2: 199–221.

Fevre, R., Rees, G. and Gorard, S. (1999) 'Some sociological alternatives to human capital theory', *Journal of Education and Work*, 12, 2: 117–40.

Fitzpatrick, T. (2003) 'New technologies and social policy', *Critical Social Policy*, 23, 2: 131–8.

Flyvbjerg, B. (2001) *Making Social Science Matter: Why Social Inquiry Fails and how it can Succeed Again*, Cambridge: Cambridge University Press.

Forman, C., Goldfarb, A. and Greenstein, S. (2002) *Digital Dispersion: An Industrial and Geographic Census of Commercial Internet Use*, Cambridge, MA: National Bureau of Economic Research.

Frank, M., Reich, N. and Humpreys, K. (2003) 'Respecting the human needs of students in the development of e-learning', *Computers and Education*, 40: 57–70.

Friesen, N. and Anderson, T. (2004) 'Interaction for lifelong learning', *British Journal of Educational Technology*, 35, 6: 679–87.

Frith, S. (1983) 'The pleasures of the hearth', in J. Donald (ed.) *Formations of Pleasure*, London: Routledge.

Fryer, R. (1997) *'Learning for the twenty-first century'* London: Department of Education and Employment.

Furlong, J., Facer, K., Sutherland, R. and Furlong, R. (2001) 'A curriculum without walls' *Cambridge Journal of Education*, 30, 1: 91–110.

Furlong, J., Selwyn, N. and Gorard, S. (2001) 'Adult Learning @ Home: lifelong learning and the ICT revolution – a project outline', paper presented to BERA 2001, Leeds.

Further Education Unit (FEU) (1993) *Paying their Way: The Experiences of Adult Learners in Vocational Education and Training in FE Colleges*, London: Further Education Unit.

Gadamer, H. (2001) 'Education is self-education' *Journal of Philosophy of Education*, 35, 4: 529–38.

Garnham, N. (1997) 'Amartya Sen's "capabilities" approach to the evaluation of welfare: its application to communication', *Javnost – The Public*, 4, 4: 25–34.

Garsten, C. and Wulff, H. (2003) *New Technologies at Work: People, Screens and Social Virtuality*, Oxford: Berg.

Giacquita, J., Bauer, J. and Levin, J. (1993) *Beyond Technology's Promise*, Cambridge: Cambridge University Press.

Gibson, C. (2003) 'Digital divides in New South Wales: a research note on socio-spatial inequality using 2001 Census data on computer and internet technology', *Australian Geographer*, 34, 2: 239–57.

Girod, R. (1990) *Problems of Sociology in Education*, Paris: UNESCO.

Glennan, T. and Melmed, A. (1996) *Fostering the Use of Educational Technology: Elements of a National Strategy*, Santa Monica, CA: Rand.

Gorard, S. and Rees, G. (2002) *Creating a Learning Society?*, Bristol: Policy Press.

Gorard, S. (2002) 'The role of causal models in education as a social science', *Evaluation and Research in Education*, 16, 1: 51–65.

Gorard, S. (2003) 'Lifelong learning trajectories in Wales: results of the NIACE Adults Learners Survey', in F. Aldridge and N. Sargant (eds) *Adult Learning and Social Division: Volume 2*, Leicester: NIACE.

Gorard, S. (2004) *Combining Methods in Educational and Social Research*, London: Open University Press.

Gorard, S., Rees, G. and Fevre, R. (1999) 'Two dimensions of time: the changing social context of lifelong learning', *Studies in the Education of Adults*, 31, 1: 35–48.

Gorard, S., Selwyn, N. and Madden, L. (2003) 'Logged on to learning? Assessing the impact of technology on participation in lifelong learning', *International Journal of Lifelong Learning*, 22, 3: 281–96.

Graham, S. (2002) 'Building urban digital divides? Urban polarisation and information and communication technologies', *Urban Studies*, 39, 1: 33–56.

Graham, S. (2004) 'Beyond the "dazzling light": from the dreams of transcendence to the 'remediation' of urban life', *New Media and Society*, 6, 1: 16–25.

Green, A. (2001) 'Unemployment, nonemployment and labour-market disadvantage', *Environment and Planning A*, 33: 1361–4.

Greenstein, S. (1999) 'On the net: on the recent commercialisation of access infrastructure', *Information Impacts*, December (www.cisp.org/imp/december_99).

Habib, L. and Cornford, T. (2001) 'Computers in the home: domestic technology and the process of domestication', in J. Podlogar and S. Avgerinou (eds) *Proceedings of the 9th European Conference on Information Systems (ECIS)*, Bled, Slovenia, 27–29 June, Bled: University of Maribor.

Habib, L. and Cornford, T. (2002) 'Computers in the home: domestication and gender', *Information Technology and People*, 15, 2: 159–74.

Haddon, L. (1988) 'The home computer: the making of the consumer electronic', *Science as Culture*, 2: 7–51.

Haddon, L. and Silverstone, R. (1994) 'The careers of information and communication technologies in the home', in K. Bjerg and K. Borreby (eds) *Proceedings of the International Working Conference on Home Oriented Informatics, Telematics and Automation*', Copenhagen, 27 June–1 July, Copenhagen: University of Copenhagen.

Hager, P. (2000) 'Know-how and workplace practical judgement', *Journal of Philosophy of Education* 34, 2: 281–96.

Hand, A., Gambles, J. and Cooper, E. (1994) *Individual Commitment to Learning: Individuals, Decision-making about Lifetime Learning*, London: Employment Department.

Hand, J. (2005) 'The people's network? Self-education and empowerment in the public library', *Information, Communication and Society* (in press).

Hara, N. and Kling, R. (2002) 'Students' difficulties in a web-based distance education course: an ethnographic study', in W. Dutton and B. Loader (eds) *Digital Academe: New Media and Institutions in Higher Education and Learning*, London: Taylor & Francis/ Routledge.

Hargittai, E. (2004) 'Internet access and use in context', *New Media and Society*, 6, 1: 115–21.

Harrison, R. (1993) 'Disaffection and access', in J. Calder (ed.) *Disaffection and Diversity: Overcoming Barriers to Adult Learning*, London: Falmer Press.

Hasluck, C. and Green, A. (1998) *Living in Hackney and Islington: Labour Market Experience and Working Lives*, Warwick: University of Warwick, Institute for Employment Research.

Hawkey, R. (2002) 'The lifelong learning game?', *Computers and Education*, 38: 5–20.

Hedra Consultancy (2002) *One Billion Pound Government Internet Strategy Misses Target*, Press Release, 29 December, Farnham: Hedra Consultancy.

Heller, F. (1987) 'The technological imperative and the quality of employment', *New Technology, Work and Employment*, 2: 19–26.

Henman, P. and Adler, M. (2003) 'Information technology and the governance of social security', *Critical Social Policy*, 23, 2: 139–65.

Hodkinson, P., Colley, H. and Malcolm, J. (2003) 'Dimensions of formality and informality in 'informal' learning settings', paper presented to the CRLL Second International Conference, Glasgow, June.

Hook, S. (2003) 'Basic skills are way off course', *Times Educational Supplement FE Focus*, 26 September: 1.

Hook, S. (2004a) 'Cash dries up for illiterate adults', *Times Educational Supplement FE Focus*, 26 March: 1.

Hook, S. (2004b) 'Distance learning misses the mark', *Times Educational Supplement FE Focus*, 12 March: 1.

Hudson, J. (2003) 'E-Galitarianism? The information society and New Labour's repositioning of welfare', *Critical Social Policy*, 23, 2: 268–90.

Hughes, C. (2002) 'Beyond the poststructuralist-modern impasse: the woman returner as "exile" and "nomad"', *Gender and Education*, 14, 4: 411–24.

Hutton, W. (2004) 'Why politics must connect', *The Observer*, 1 August: 26.

ICM (2004) 'Our friends electric: Spark/Toyota Prius ICM Poll', *The Guardian*, *Spark* supplement, May: 26–9.

Imel, S. (2001) 'Learning technologies in adult education', *Myths and Realities*, 17: 1–2.

Jackson, S. (2003) 'Lifelong earning: working-class women and lifelong learning', *Gender and Education*, 15, 4: 365–76.

Jenkins, A., Vignoles, A., Wolf, A. and Galindo-Rueda, F. (2002) *The Determinants and Effects of Lifelong Learning*, Working Paper, Centre for the Economics of Education, London School of Economics.

Jeris, L. (2002) 'Comparison of power relations within electronic and face-to-face classroom discussions: a case study', *Australian Journal of Adult Learning*, 42, 3: 300–11.

Jordan, A. (1991) 'Social class, temporal orientation and mass media use within the family system', *Critical Studies in Mass Communication*, 9: 374–86.

Jordan, A. (2003) 'A family systems approach to examining the role of the internet in the home', in J. Turow and A. Kavanaugh (eds) *The Wired Homestead: An MIT Sourcebook on the Internet and the Family*, Cambridge, MA: MIT Press.

Jung, J., Qiu, J. and Kim, Y. (2001) 'Internet connectedness and inequality: beyond the divide', *Communication Research*, 28, 4: 507–35.

Katz, J. and Aakhus, M. (2002) *Perpetual Contact*, Cambridge: Cambridge University Press.

Kelly, T. (1992) *A History of Adult Education in Great Britain*, Liverpool: Liverpool University Press.

Kennedy, H. (1997) *Learning Works: Widening Participation in Further Education*, Coventry: Further Education Funding Council.

Kennedy-Wallace, G. (2002) 'E-learning is booming but the UK still lags behind', *The Guardian* Education supplement, 16 April: 49.

Kenway, J., Bigum, C., Fitzclarence, L., Collier, J. and Tregenza, K. (1994) 'New education in new times', *Journal of Education Policy*, 9, 4: 317–33.

Kingston, P. (2004a) 'Adults learning less under Labour', *The Guardian*, Education supplement, 18 May: 15.

Kingston, P. (2004b) 'Actions speak louder than words', *The Guardian*, Education supplement, 9 November: 14.

La Valle, I. and Blake, M. (2001) *National Adult Learning Survey 2001 Research Report 321*, Nottingham: Department for Education and Science.

Lægran, A. and Stewart, J. (2003) 'Nerdy, trendy or healthy? Configuring the internet café', *New Media and Society*, 5, 3: 357–77.

Lally, E. (2002) *At Home with Computers*, Oxford: Berg.

Lamb. J. (2004) 'A personal map of cyberspace', *The Guardian*, Society supplement, 24 March: 9.

Lash, S. (2002) *Critique of Information*, London: Sage.

Law, J. and Hassard, J. (1999) *Actor-network Theory and After*, Oxford: Blackwell.

Leadbetter, C. (1999) *Living on Thin Air*, Harmondsworth: Penguin.

Learning and Skills Council (LSC DELG) (2002) *Report of the Learning and Skills Council's Distributed and Electronic Learning Group*, London: Learning and Skills Council.

Lehtonen, T. and Sundell, R. (2004) 'The domestication of new technologies as a set of trials' (http://www.valt.helsinki.fi/comm/argo/turokim.htm).

Lenhart, A., Simon, M. and Graziano, M. (2001) *The Internet and Education: Findings of the Pew Internet and American Life Project*, Washington, DC: Pew Internet and American Life Project.

Lenhart, A., Horrigan, J., Rainie, L., Allen ,K., Boyce, A., Madden, M. and O'Grady, E. (2003) *The Ever-shifting Internet Population: A New Look at Internet Access and the Digital Divide*, Washington, DC: Pew Internet and American Life Project.

Lewis, I. (2002) *Ask Ivan: Live Webchat event with Ivan Lewis MP, Parliamentary Under-Secretary of State for Adult Learning and Skills*, 26 November (http://www.dfes.gov.uk).

Lewis, L. and Delcourt, M. (1998) 'Adult basic education students' attitudes toward computers', in Standing Conference on University Teaching and Research in the Education of Adults (SCUTREA), *Proceedings of 1998 Conference*: 238–42.

Liff, S., Steward, F. and Watts, P. (2002) 'New public places for internet access: networks for practice-based learning and social inclusion', in S. Woolgar (ed.) *Virtual Society? Technology, Cyberbole, Reality*, Oxford: Oxford University Press.

Lim, C. (2002) 'A theoretical framework for the study of ICT in schools: a proposal', *British Journal of Educational Technology*, 33, 4: 411–21.

Limb, A. (2003) talk given to 'E-learning and post-16 education: the present and potential role of ICT', Seminar, IPPR/Design Council, London: 28 May.

Livingstone, D. (2000) 'Researching expanded notions of learning and work and underemployment: findings of the first Canadian survey of informal learning practices', *International Review of Education*, 46, 6: 491–514.

Livingstone, S., Holden, K. and Bovill, M. (1999) 'Children's changing media environment', in C. von Feilitzen and U. Carlsson (eds) *Children and Media: Image, Education, Participation*, Göteborg: Nordicom/ UNESCO.

Loges, W. and Jung, J. (2001) 'Exploring the digital divide: Internet connectedness and age', *Communication Research*, 28, 4: 536–62.

Lohman, M. (2000) 'Environmental inhibitors to informal learning in the workplace: a case study of public school teachers', *Adult Education Quarterly*, 50, 2: 83–101.

Lovin, B. (1992) 'Professional learning through workplace partnerships', in H. Baskett and V. Marsick (eds) *Professionals' Ways of Knowing*, San Francisco, CA: Jossey-Bass.

Lundvall, B. and Johnson, B. (1994) 'The learning economy', *Journal of Industrial Studies*, 1, 2: 23–42.

Lupton, D. and Noble, G. (2002) 'Mine/not mine: appropriating personal computers in the academic workplace', *Journal of Sociology*, 38, 1: 5–23.

Lyon, D. (1988) *The Information Society: Issues and Illusions*, Cambridge: Polity.

Lyon, D. (1995) 'The roots of the information society idea', in N. Heap, R. Thomas, G. Einon, R. Mason, R. and H. Mackay (eds) *Information Technology and Society: A Reader*, London: Sage.

Macleod, D. (2003) *Widening Adult Participation: A Review of Research and Development*, London: Learning and Skills Development Agency.

Maddux, C. (1989) 'The harmful effects of excessive optimism in educational computing', *Educational Technology*, 29, 4: 23–9.

Maguire, M., Maguire, S. and Felstead, A. (1993) *Factors Influencing Individual Commitment to Lifetime Learning. A Literature Review*, Leicester: Employment Department.

Maki, R., Maki, W., Patterson, M., and Whittaker, P. (2000) 'Evaluation of a web-based introductory psychology course: I. Learning and satisfaction in on-line versus lecture courses', *Behaviour Research Methods, Instruments, and Computers*, 32: 230–9.

Mallet, S. (2004) 'Understanding home: a critical review of the literature', *The Sociological Review*, 52, 1: 62–89.

Mannheim, K. (1985) *Ideology and Utopia*, London: Harcourt.

Margetts, H. and Dunleavy, P. (2002) *Better Public Services Through e-Government*, London: National Audit Office.

Marshall, P. (1997) 'Technophobia: video games, computer hacks and cybernetics', *Media International Australia*, 85: 70–8.

Marsick, V. and Watkins, K. (1990) *Informal and Incidental Learning in the Workplace*, London: Routledge.

Masuda, Y. (1981) *The Information Society as Post-industrial Society*, Bethesda, MD: World Futures Society.

Mathieson, S. (2003) 'Digitally divided by choice', *The Guardian*, Online supplement, 18 September: 1–2.

Mayes, T. (2000) 'Pedagogy, lifelong learning and ICT', in Scottish Forum on Lifelong Learning, *Role of ICT in Supporting Lifelong Learning*, Stirling: Centre for Research in Lifelong Learning, University of Stirling.

McAdams, D. (1998) 'The role of defence in the life story', *Journal of Personality*, 66: 1125–46.

McGivney, V. (1993) 'Participation and non-participation: a review of the literature', in R. Edwards, S. Sieminski and D. Zeldin (eds) *Adult Learners, Education and Training*, London: Routledge.

McIntosh, N. (2004) 'Homing in on the future', *The Guardian*, Online supplement, 15 January: 28.

Menchik, D. (2004) 'Placing cybereducation in the UK classroom', *British Journal of Sociology of Education*, 25, 2: 193–213.

Merton, R. (1976) *Sociological Ambivalence and Other Essays*, New York: Free Press.

Miles, I. (1996) 'The information society: competing perspectives on the social and economic implications of information and communications technologies', in W. Dutton (ed.) *Information and Communications Technologies: Visions and Realities*, Oxford: Oxford University Press.

Miller, D. (1994) 'The young and the restless in Trinidad: a case of the local and the global in mass consumption', in R. Silverstone and E. Hirsch (eds) *Consuming Technologies: Media in Information and Domestic Spaces*, London: Routledge.

Mitchell, W. (2001) 'Multilevel modelling might not be the answer', *Environment and Planning A*, 33: 1357–60.

Moore, N. (1998) 'Confucius or capitalism? Policies for an information society', in B. Loader (ed.) *Cyberspace Divide: Equality, Agency and Policy in the Information Society*, London: Routledge.

Moran-Ellis, J. and Cooper, G. (2000) 'Making connections: children, technology and the national grid for learning', *Sociological Research Online*, 5, 3 (http://www.socresonline.org.uk).

Morley D. (2003) 'What's "home" got to do with it? Contradictory dynamics in the domestication of technology and the dislocation of domesticity', *European Journal of Cultural Studies*, 6, 4: 435–58.

Morley, D. and Silverstone, R. (1990) 'Domestic communication: technologies and meanings', *Media, Culture and Society*, 12, 1: 31–55.

Moser Group (1999) *A Fresh Start: Improving Numeracy and Literacy*, London: Basic Skills Agency.

Mossberger, K., Tolbert, C. and Stansbury, M. (2003) *Virtual Inequality: Beyond the Digital Divide*, Washington, DC: Georgetown University Press.

Murdock, G. (2002) 'Debating digital divides', *European Journal of Communication*, 17, 3: 385–90.

Murdock, G. (2004a) 'Past the posts: rethinking change, retrieving critique', *European Journal of Communication*, 19, 1: 19–38.

Murdock, G. (2004b) 'Rethinking the dynamics of exclusion and participation?', paper given to OfCOM Consumer Panel Seminar, London: 29 November.

Murdock, G., Hartmann, P. and Gray, P. (1992) 'Contextualising home computing: resources and practices', in R. Silverstone and E. Hirsch (eds) *Consuming Technologies*, London: Routledge.

Murdock, G., Hartmann, P. and Gray, P. (1996) 'Conceptualising home computing: resources and practices', in N. Heap, R. Thomas, G. Einon, R. Mason and H. Mackay (eds) *Information Technology and Society*, London: Sage.

National Center for Education Statistic (NCES, US Department of Education) (2002a) *Participation Trends and Patterns in Adult Education: 1991 to 1999*, Washington, DC: NCES.

National Center for Education Statistic (NCES, US Department of Education) (2002b) *Programs for Adults in Public Library Outlets*, Washington, DC: NCES.

National Institute for Adult and Continuing Education (NIACE) (1994) *Widening Participation: Routes to a Learning Society*, NIACE Policy Discussion Paper, Leicester: NIACE.

National Institute for Adult and Continuing Education (NIACE) (2003) *Adults Learning Survey, 2003*, Leicester: NIACE.

National Research Council (2002) *Scientific Research in Education*, Washington, DC: National Academy Press.

National Telecommunication and Information Administration (NTIA) (1995) *Falling Through the Net: A Survey of the 'Have Nots' in Rural and Urban America*, Washington, DC: NTIA (http://www.ntia.doc.gov/ntiahome/digitaldivide).

National Telecommunication and Information Administration (NTIA) (1999) *Falling Through the Net: Defining the Digital Divide'* Washington, DC: NTIA (http://www.ntia.doc.gov/ntiahome/digitaldivide).

National Telecommunication and Information Administration (NTIA) (2000) *Falling Through the Net: Toward Digital Inclusion*, Washington, DC: NTIA (http://www.ntia.doc.gov/ntiahome/digitaldivide).

Navarra, D. and Cornford, T. (2003) 'A policy making view of e-government innovations in public governance', paper presented to Americas Conference on Information Systems, Tampa, Florida, 4–6 August.

Negroponte, N. (1995) *Being Digital*, London: Coronet.

Neice, D. (1998) *Measures of Participation in the Digital Techno-structure: Internet Access*, ACTS/FAIR Working Paper No. 44, Brighton: Science Policy Research Unit.

Nettleton, S. and Burrows, R. (2003) 'E-Scaped medicine? Information, reflexivity and health', *Critical Social Policy*, 23, 2: 165–85.

Nora, S. and Minc, A. (1980) *The Computerisation of Society*, Cambridge, MA: MIT Press.

Norris, P. (2001) *Digital Divide: Civic Engagement, Information Poverty and the Internet World-wide*, Cambridge: Cambridge University Press.

Ocak, M. (2004) 'Adult learners' attitudes towards computers', paper given to the 16 Annual Ethnography and Qualitative Research on Education Conference, Albany, NY, 2–4 June, Albany, NY: SUNY.

Office of Vocational and Adult Education (OVAE, US Department of Education) (2000) *Adult Education Facts at a Glance'* Washington, DC: OVAE.

Olsen, H. (2004) 'Locals left out in the cold', *The Guardian*, e-public supplement, 26 May: 18.

O'Neil, D. (2002) 'Assessing community informatics: a review of methodological approaches for evaluating community networks and community technology centres', *Internet Research: Electronic Networking Applications and Policy*, 12, 1: 76–102.

Oppenheim, C. (1998) *An Inclusive Society: Strategies for Tackling Poverty*, London: Institute for Public Policy Research.

Organisation of Economic Co-operation and Development (OECD) (1996) *Education and training: Learning and work in a society in flux*, Paris: OECD.

Organisation of Economic Co-operation and Development (OECD) (2003) *Beyond Rhetoric: Adult Learning Policies and Practices*, Paris: OECD.

Osborne, D. and Gaebler, T. (1992) *Reinventing Government: How the Entrepreneurial Spirit is Transforming the Public Sector*, London: Penguin.

Otterson, E. (2004) 'Lifelong learning and challenges posed to European labour markets', *European Journal of Education*, 39, 2: 151–7.

Owerwien, B. (2000) 'Informal learning and the role of social movements', *International Review of Education*, 46, 6: 621–40.

Pantzar, E. (2001) 'European perspectives on lifelong learning environments in the informational society', in E. Karvonen (ed.) *Informational Societies: Understanding the Third Industrial Revolution*, Tampare: Tampare University Press.

Park, A. (1994) *Individual Commitment to Lifelong Learning: Individual's Attitudes Report on the Quantitative survey*, London: Employment Department.

Pascual Leone, J. (1998) 'Abstraction, the will, the self and modes of learning in adulthood', in M. Cecil Smith and T. Pourchot (eds) *Adult Learning and Development*, Mahwah, NJ: Lawrence Erlbaum.

Paton, N. (2003) 'Hard lessons from the big e-learning experiment', *The Guardian*, 30 August: 20–21.

Pérez Cereijo, M., Tyler-Wood, T. and Young, J. (2002) 'Student perceptions of online synchronous courses', *Proceedings of E-Learn 2000 World Conference on e-Learning in Corporate, Government, Healthcare and Higher Education*, Montreal, October.

Pinder, A. (2001) Keynote talk to the GC2001 – the Government Computing Conference, London: 1 May.

Policy Research Institute (2002) *Wired up Communities Programme: Second Interim Report*, London: Policy Research Institute.

Porter, T. (1986) *The Rise of Statistical Thinking*, Princeton: Princeton University Press.

Postman, N. (1993) *Technopoly: The Surrender of Culture to Technology*, New York: Vintage Books.

Powell, M. (2001) Speech cited in the *Washington Post*, 18 June 2001.

Public Accounts Committee (2002) *Improving Public Services Through e-Government: Public Accounts: Fifty-Fourth Report*, London: The Stationery Office.

Pyke, N. (1996) 'Dearing champions the young no-hopers', *Times Educational Supplement*, 29 March: 1.

Quibria, M., Ahmed, S., Tschang, T. and Reyes-Macasaquit, M. (2002) *Digital Divide: Determinants and Policies with Special Reference to Asia*, Manila: Asian Development Bank.

Quicke, J. (1997) 'Reflexivity, community and education for the learning society', *Curriculum Studies*, 5, 2: 139–61.

Qvortrup, L. (1984) *The Social Significance of Telematics: An Essay on the Information Society* (Philip Edmonds, trans.), Philadelphia, PA: John Benjamins.

Rainie, L. and Bell, P. (2004) 'The numbers that count', *New Media and Society*, 6, 1: 44–54.

Reddick, A. (2000) *The Dual Digital Divide: The Information Highway in Canada*, Ottawa: Public Interest Advocacy Centre.

Resnick, P. (2002) 'Beyond bowling together: sociotechnical capital', in J. Carroll (ed.) *HCI in the New Millennium*, Boston, MA: Addison-Wesley.

Rifkin, J. (2000) *The Age of Access: How the Shift from Ownership to Access is Transforming Modern Life*, London: Penguin.

Robertson, D. (1998) 'The university for industry: a flagship for demand-led training or another doomed supply-side intervention?', *Journal of Education and Work*, 11, 1: 5–22.

Rogers, E. (1995) *Diffusion of Innovations*, 2nd edn, New York: The Free Press.

Rogers, E. and Shoemaker, F. (1971) *Communication of Innovations*, New York: Free Press.

Romanyshyn, R. (1989) *Technology as Symptom and Dream*, London: Routledge.

Rosen, D. (1998) 'Using electronic technology in adult literacy education', in NCSALL (ed.) *The Annual Review of Adult Learning and Literacy*, vol. 1, Cambridge, MA: National Center for the Study of Adult Learning and Literacy.

San-Segundo, M. and Valiente, A. (2003) 'Family background and returns to schooling in Spain', *Education Economics*, 11, 1: 39–52.

Sargant, N. (2000) *The Learning Divide Revisited*, Leicester: NIACE.

Sargant, N. and Aldridge, F. (2002) *Adult Learning and Social Division: A Persistent Pattern*, vol. 1, Leicester: NIACE.

Sarojni, C., McNickle, C. and Berwyn, C. (2002) *Learner Expectations and Experiences: An Examination of Student Views of Support in Online Learning*, Leabrook, South Australia: National Centre for Vocational Education Research.

Sawchuk, P. (2003) *Adult Learning and Technology in Working-class Life*, Cambridge: Cambridge University Press.

Sawhney, H. (2000) 'Universal access: separating the grain of truth from the proverbial chaff', *The Information Society*, 16, 2: 161–4.

Schattschneider, E. (1960) *The Semi-sovereign People*, New York: Holt, Rinehart and Winston.

Schofield-Clark, L., Demont-Heinrich, C. and Webber, S. (2004) 'Ethnographic interviews on the digital divide', *New Media and Society*, 6, 4: 529–47.

Schuller, T., Preston, J., Hammond, C., Brassett-Grundy, A. and Bynner, J. (2004) *The Benefits of Learning*, London: RoutledgeFalmer.

Schwartz, R. and Duvall C. (2000) 'Distance education: relationship between academic performance and technology-adept adult students', *Education and Information Technologies*, 5, 3: 177–87.

Selwyn, N. (2000) 'Researching computers and education: glimpses of the wider picture', *Computers and Education*, 34, 2: 93–101.

Selwyn, N. (2002) *Telling tales on Technology*, Aldershot: Ashgate.

Selwyn, N. and Gorard, S. (2002) *The Information Age: Technology, Learning and Exclusion*, Cardiff: University of Wales Press.

Servon, L. and Nelson, M. (2001) 'Community technology centres and the urban technology gap', *International Journal of Urban and Regional Research*, 25, 2: 419–26.

Shearman, C. (1999) *Local Connections: Making the Net Work for Neighbourhood Renewal*, London: Communities Online.

Shields, R. (2004) 'Governing in the knowledge economy', seminar given to Cardiff University School of Social Sciences, Cardiff, April.

Shields, R. (2003) *The Virtual*, London: Routledge.

Shuklina, E. (2001) 'The technologies of self education', *Russian Education and Society*, 43, 2: 57–78.

Silverstone, R. (1993) 'Time, information and communication technologies and the household', *Time and Society*, 2, 3: 283–311.

Silverstone, R. (1996) 'Future imperfect: information and communication technologies in everyday life', in W. Dutton (ed.) *Information and Communications Technologies: Visions and Realities*, Oxford: Oxford University Press.

Silverstone, R. and Haddon, L. (1996) 'Design and the domestication of information and communications technologies: technical change and everyday life', in R. Mansell and R. Silverstone (eds) *Communication by Design: The Politics of Communication Technologies*, Oxford: Oxford University Press.

Silverstone, R. and Hirsch, E. (1992) *Consuming Technologies: Media and Information in Domestic Spaces*, London: Routledge.

Silverstone, R., Hirsch, E. and Morley, D. (1992) 'Information and communication technologies and the moral economy of the household', in R. Silverstone and E. Hirsch (eds) *Consuming Technologies: Media and Information in Domestic Spaces*, London: Routledge.

Smet, K., Roe, K. and Van Rompaey, V. (2002) 'Do children make a difference? The media structure in households with and without children', IAMCR Conference, Barcelona, 21–26 July.

Smith, I. (1999) 'What do we know about public library use?', *ASLIB Proceedings*, 51, 9: 302–14.

Southwest Educational Development Laboratory (SEDL) (1995) *Networks for Goals 2000 Reform: Bringing the Internet to K-12 Schools*, Austin, TX: Southwest Educational Development Laboratory.

Starr, S. (2002) 'Tenants' associations', *Spiked*, 23 December (www.spiked-online.com).

St Clair, R. (2004) 'A beautiful friendship? The relationship of research to practice in adult education', *Adult Education Quarterly*, 54, 3: 224–41.

Stewart, J. (1999) 'Cybercafes: computers in the community, not communities in the computers', Social Learning Multimedia (SLIM) Project, University of Edinburgh, mimeo.

Stewart, J. (2002) 'Encounters with the information society', unpublished PhD thesis, University of Edinburgh.

Strover, S. (2003) 'Remapping the digital divide', *The Information Society*, 19: 275–7.

Strover, S., Chapman, G. and Waters, J. (2004) 'Beyond community networking and CTCs: access, development and public policy', *Telecommunications Policy*, 28: 465–85.

Struys, K., Roe, K. and Van Rompaey, V. (2001) 'Children's influence on the family purchase of media', paper given to International Communication Association Conference, Washington, DC, 24–28 May.

Sutherland, J. (2004) 'America is addicted to the internet, burgers and drugs', *The Guardian*, G2 supplement, 6 December: 7.

Sutton, B. (2003) Talk given to 'E-learning and post-16 education: the present and potential role of ICT', Seminar, IPPR/Design Council, London, 28 May.

Tally, B. (2004) 'Talking about (school and) technology: data collection of student/user perspectives', paper presented to American Education Research Association Conference, San Diego, April.

Tambini, D. (2000) *Universal Internet Access: A Realistic View*, London: Institute for Public Policy Research.

Tasker, M. and Packham, D. (1993) 'Industry and higher education: a question of values', *Studies in Higher Education*, 18, 2: 127–36.

Taylor, S. and Spencer, L. (1994) *Individual Commitment to Lifelong Learning: Individual's Attitudes*, London: Employment Department.

The Economist (2001) 'Getting better all the time: a survey of technology and development', *The Economist*, supplement 10 November.

The Guardian (2004) 'Poll shows new bond with PCs', *The Guardian*, 24 February: 8.

Tiffin, J. and Rajasingham, L. (1995) *In Search of the Virtual Class: Education in an Information Society*, London: Routledge.

Titmus, C. (1994) 'The scope and characteristics of educational provision for adults', in J. Calder (ed.) *Disaffection and Diversity. Overcoming Barriers to Adult Learning*, London: Falmer Press.

Todd, M. and Tedd, L. (2000) 'Training courses for ICT as part of lifelong learning in public libraries: experiences with a pilot scheme in Belfast Public Libraries', *Program*, 34, 4: 375–83.

Tolmie, A. and Boyle, J. (2000) 'Factors influencing the success of computer mediated communication environments in university teaching: a review and case study', *Computers and Education*, 34, 2: 119–40.

Tough, A. (1978) 'Major learning efforts: recent research and future directions', *Adult Education Quarterly*, 28, 4: 250–63.

Toulouse, C. (1997) 'Introduction', in C. Toulouse and T. Luke (eds) *The Politics of Cyberspace*, London: Routledge.

Touraine, A. (1969) *La société post-industrielle. Naissance d'une société*, Paris: Denoël.

Trow, M. (1999) 'Lifelong learning through the new information technologies', *Higher Education Policy*, 12, 2: 201–17.

Tuomi, I. (2000) 'Beyond the digital divide', paper presented to the UCB Human-Centric Computing Workshop, University of California, Berkeley, July.

University of California, Los Angeles (UCLA) (2000) *UCLA Internet Report: Surveying the Digital Future'*, Los Angeles, CA: UCLA Centre for Communication Policy.

Uotinen, J. (2003) 'Involvement in the information society: the Joensuu community resource centre netcafé', *New Media and Society*, 5, 3: 335–56.

Valentine, G. and Holloway, S. (1999) 'The vision thing: schools and information and communication technology', *Convergence* 5: 63–79.

Van den Berg, L. and van Winden, W. (2002) *Information and Communication Technology as Potential Catalyst for Sustainable Urban Development*, Aldershot: Ashgate.

Van Rompaey, V., Roe, K. and Struys, K. (2002) 'Children's influence on internet access at home. Adoption and use in the family context', *Information, Communication and Society*, 5, 2: 189–206.

Van Zoonen, L (2002) 'Gendering the internet: claims, controversies and cultures', *European Journal of Communication*, 17, 1: 5–23.

Volti, R. (1992) *Society and Technological Change*, New York: St Martin's Press.

Vryzas, K. and Tsitouridou, M. (2002) 'The home computer in children's everyday life: the case of Greece', *Journal of Educational Media*, 27, 1/2: 9–18.

Wakeford, N. (2003) 'The embedding of local culture in global communication: independent internet cafés in London', *New Media and Society*, 5, 3: 379–99.

Walker, A. (1997) 'Introduction', in A. Walker and C. Walker (eds) *Britain Divided: The Growth of Social Exclusion in the 1980s and 1990s*, London: Child Action Poverty Group.

Walker, D. (2004) 'Opportunity knocks', *Times Educational Supplement – Online supplement*, 18 June: 10–11.

Walker, L., Matthew, B. and Black, F. (2004) 'Widening access and student non-completion: an inevitable link?', *International Journal of Lifelong Education*, 23, 1: 43–59.

Warren, S. (2004) 'Review article: "Creating a learning society? The information age: world class schools"', *Journal of Education Policy*, 19, 1: 105–9.

Webster, F. (2001) 'Global challenges and national answers', in E. Karvonen (ed.) *Informational Societies: Understanding the Third Industrial Revolution*, Tampare: Tampare University Press.

Webster, F. (2002) 'Cybernetic life: limits to choice', in J. Armitage and J. Roberts (eds) *Living with Cyberspace: Technology and Society in the 21st Century*, London: Continuum.

Weiland, S. (1995) 'Review of "the narrative study of lives"', *Qualitative Studies in Education*, 8, 1: 99–105.

Weinburg, A. (1966) 'Can technology replace social engineering?', reprinted in G. Hawisher and C. Selfe (eds) *Literacy, Technology and Society*, New York: Prentice Hall.

Weingardt, K. (2000) 'Viewing ambivalence from a sociological perspective: implications for psychotherapists', *Psychotherapy* 37, 4: 298–306.

Wellman, B. (2001) 'Computer networks as social networks', *Science*, 293, 14 September: 2031–4.

Wellman, B. (2004) 'The three stages of internet studies', *New Media and Society*, 6, 1: 123–9.

Wessel, W. (2000) 'Technology in the classroom: implications for teacher education', in D. Willis, J. Price and J. Willis (eds) *Society for Information Technology and Teacher Education International Conference: Proceedings of SITE 2000*, Charlottesville, VA: Association for the Advancement of Computing in Education.

Wheelock, J. (1992) 'Personal computers, gender and an institutional model of the household', in R. Silverstone and E. Hirsch (eds) *Consuming Technologies*, London: Routledge.

Whittaker, M. (2003) 'Saved from the new illiteracy', *Times Educational Supplement, Work-related Learning*, 3 October: 8–9.

Wilhelm, A. (2000) *Democracy in the Digital Age*, London: Routledge.

Williams, A. and Alkalimat, A. (2004) 'A census of public computing in Toledo, Ohio', in D. Schuler and P. Day (eds) *Shaping the Network Society: The New Role of Civic Society in Cyberspace*, Cambridge, MA: MIT Press.

Williams, R. (1999) *The National Appropriation of Multimedia: Introduction to the National Studies Conducted under the EC TSER Project: Social Learning in Multimedia*, Edinburgh: University of Edinburgh, Research Centre for Social Studies.

Williamson, D. (2003) 'Lessons planned for your local', *Western Mail*, 10 April: 25.

Wilson, B. and Lowry, M. (2000) 'Constructivist learning happens all the time on the web', in E. Burge (ed.) *Learning Technologies: Reflective and Strategic Thinking, New Directions for Adult and Continuing Education No. 88*, San Francisco, CA: Jossey-Bass.

Wilson, E. (2000) *Closing the Digital Divide: An Initial Review. Briefing the President*, Washington, DC: Internet Policy Institute.

Wilson, P. (1973) 'Situational relevance', *Information Storage and Retrieval*, 9: 457–71.

Winner, L. (1980) 'Do artefacts have politics?', *Daedalus*, 109: 121–36.

Wise, J.M. (1997) *Exploring Technology and Social Space*, Thousand Oaks, CA: Sage.

Woolgar, S. (1991) 'Configuring the user: the case of usability trials', in J. Law (ed.) *A Sociology of Monsters: Essays on Power, Technology and Domination*, London: Routledge.

Woolgar, S. (1996) 'Technologies as cultural artefacts', in W. Dutton (ed.) *Information and Communication Technologies*, Oxford: Oxford University Press.

World Internet Project (2003) International Conference of World Internet Project, 16–19 July, Oxford.

Wresch, W. (2004) 'Review article: the information age', *The Information Society*, 20: 71–2.

Yang, B. and Lester, D. (2000) 'Who buys their textbooks online?', *Psychological Reports*, 87, 2/3: 1183–4.

Yang, B. and Lester, D. (2002) 'Buying textbooks online', *Psychological Reports*, 91, 2/3: 1222–4.

Yang, B., Lo, P. and Lester, D. (2003) 'Purchasing textbooks online', *Applied Economics*, 35, 11: 1265–9.

Zborovskii, G. and Shuklina, E. (2001) 'Self education as a sociological problem', *Russian Education and Society*, 40, 10: 65–91.

Author index

Subject index